CAMBRIDGE COUNTY GEOGRAPHIES

SCOTLAND

General Editor: W. MURISON, M.A.

T0364307

FORFARSHIRE

Cambridge County Geographies

FORFARSHIRE

by

EASTON S. VALENTINE, M.A.

Headmaster of the English Department in Dundee High School :
Formerly Examiner in English in the University of Glasgow
and in the University of St Andrews

With Maps, Diagrams and Illustrations

Cambridge :
at the University Press
1912

CAMBRIDGE UNIVERSITY PRESS
Cambridge, New York, Melbourne, Madrid, Cape Town,
Singapore, São Paulo, Delhi, Mexico City

Cambridge University Press
The Edinburgh Building, Cambridge CB2 8RU, UK

Published in the United States of America by Cambridge University Press, New York

www.cambridge.org
Information on this title: www.cambridge.org/9781107659148

© Cambridge University Press 1912

First published 1912
First paperback edition 2013

A catalogue record for this publication is available from the British Library

ISBN 978-1-107-65914-8 Paperback

CONTENTS

ILLUSTRATIONS

ILLUSTRATIONS

MAPS

The illustrations on pp. 86 and 88 are reproduced by permission of Messrs T. C. and E. C. Jack; those on pp. 93, 94, and 96 by kind permission of the Society of Antiquaries of Scotland; that on p. 58 is from the *Encyclopaedia Britannica* (11th Edition); the map facing p. 39 is reproduced by the courtesy of J. Hannay Thompson, Esq., M.I.C.E., Harbour Engineer, Dundee; the portrait on p. 141 is from the National Portrait Gallery, London; that on p. 143 is from the Scottish National Portrait Gallery, Edinburgh; the remainder of the views are from photos by Messrs J. Valentine and Sons.

*Map now available for download at www.cambridge.org/9781107659148

1. County and Shire. The County of Forfar.

Most of the divisions of Great Britain are known as shires or counties, names often, but not always, interchangeable. *Shire* comes from the Old English *scir*, meaning administration, charge. Often, though incorrectly, our early kings are represented as having divided up their realm and as having given shares of it to various noblemen to rule in subordination to the crown. Rather, we should regard many of the present divisions of the country as the relics of older kingdoms now merged in the larger unity of Great Britain.

County, or the district presided over by a *count*, is a Norman-French name and was not used on this side of the Channel until the Conquest. The title of count was regarded as the equivalent of our *earl* and was soon replaced by the latter, though the earl's lady is still called *countess*. For long the earl's duties and privileges were those of the original count. The *comes*, or companion of the king, he was the ruler under his sovereign of a county or shire, and he derived an official revenue from his earldom.

The county or shire treated of in this volume has, like a few others, two distinct names, Angus and Forfar. Angus is much the older, although Forfar is now much the commoner. Angus is held by some to have been the name of a Scottish prince who was granted the district by his father, while others interpret it as meaning a particular kind of hill. The designation Forfarshire is taken from the name of the county town.

Geographically Angus and the adjoining county of Mearns or Kincardine belong to one district, and it is thought that they formed originally one independent province, which, however, became part of the wider region of Pictavia. They appear at one time to have been ruled by a single maormor, but if so they must have been separated at an early date.

2. General Characteristics.

The high place Forfarshire takes on the list of Scottish counties is largely due to the fact that its industries are so varied. It has a long seaboard with several good harbours and valuable fishing-stations. The estuary of the Tay, though not without its dangers to navigation, has had its natural facilities carefully fostered by the harbour authorities of Dundee, and occupies the third place in Scotland for ship-building and the harbourage of vessels, being excelled only by the Forth and the Clyde.

Angus has played no inconsiderable part in the general history of the country. If one hears less of its battle-fields than of others, it was on at least one occasion the scene of a conflict—that of Nechtan's Mere—decisive in shaping the trend of subsequent events in Scotland; if it has not the fame of the Border, its proximity to the Grampians, a boundary line far more momentous than the Cheviots, as marking the limits of Celt and Saxon, laid it open to incursions quite as formidable and quite as resolutely repelled as the raids of the southron into Scotland. Its sons have made for themselves a name by both flood and field and also in the more useful if less showy pursuits of peace.

The county contains many relics that point back to the dawn of civilization in these islands: its ancient forts and camps belong to a time when history was yet unwritten. Its ruined castles and strongholds are intimately associated in song and story with the wild warfare of clannish and feudal days. Its abbeys, cathedrals, and other religious foundations are proof of its wide influence in the ecclesiastical affairs of a bygone age.

The regional characteristics of Forfarshire, ranging from the mountainous district of the Grampians in the north through the fertile plain to the alluvial and sandy shores of the Tay and the rockbound coast, with its marvellous caves between Arbroath and Montrose, provide the botanist and the geologist with an unusually varied field. Nor are its scenery and other kindred attractions any less diverse. The rocks and tarns, the corries and passes on its northern frontier attract the cragsman and the sportsman.

Its mineral wealth, though less by the absence of coal

and iron deposits than several more favoured districts of
Scotland, is very considerable, for its quarries yield a
valuable supply of building stones and paving stones.
But it is to the quality and abundance of its agricultural
produce, and still more to the extent and pre-eminence of

Masons' Cave, near Arbroath

its textile manufactures that Forfarshire owes its importance in population and industry. Strathmore, a large
part of which is within its boundaries, is one of the most
fruitful farming districts in the country; and the linen
and jute fabrics of Dundee and a whole circle of smaller
towns in Angus find a world-wide market.

3. Size. Shape. Boundaries.

Measuring direct from Blacklunnans to Scurdie Ness
near Montrose, and from Cock Cairn to Buddon Ness,
which mark the four extreme points of the modern
county, we get two lines each 36 miles long. Its area
is 559,171 acres or roughly 807 square miles, and it is
thus the eleventh in point of size amongst the counties of
Scotland. It is less than one-fifth of Inverness-shire, the
largest, and more than twenty-two times the size of
Clackmannan, the smallest. Its boundary line measures,
when zigzags are taken into account, something like
190 miles, of which about 44 are washed by the waters
of the ocean and the firth.

Forfarshire is one of the most compact of Scottish
counties, being nearly circular in shape. The landward
boundary marches with the counties of Perth, Aberdeen,
and Kincardine. To the east of the Ericht and the
Shee, Perthshire "cuts itself a monstrous cantle out";
Aberdeenshire drives a wedge deeply into its neighbour
to the south-east of Loch Muick; and in the north-east
by its adherence to the inward curve of the North Esk,
Forfar loses a crescent-shaped mass of land to Kincardine.
With these exceptions the boundary line of the county is
fairly symmetrical. The curve of the coast from Dundee
to St Cyrus is materially broken only by the promontory
of Buddon Ness, the indentation of Lunan Bay, and the
tidal waters of Montrose Basin.

The north-west corner-stone of Forfarshire, sometimes given as the huge, rugged mass of Cairn-na-Glasha

(3484 ft.), is in reality Glas Maol (3502 ft.), the highest summit of which is in the county. The boundary skirts the edge of the precipices of Creag Leacach as far as Cairn Aigha, then striking direct south it reaches in the parish of Blacklunnans its most westerly point, on the Water of Shee. After following this for a few miles it trends south-east and runs round the Perthshire Hill of

Lunan Bay

Alyth, and in a zigzag course crosses the Isla, beyond which it follows for some distance the Dean. A long tongue of the county then shoots westwards into Perth-shire so as to include the parish of Kettins. From this point to the Tay at Invergowrie, the line, which once again projects into Perthshire around the parish of Benvie, is straggling and artificial.

The main range of the Grampians with its lofty

but rounded summits divides Forfar on the north from
Aberdeen. The principal heights are Tolmount, Cairn
Bannoch, Broad Cairn, Dog Hillock, Black Head of
Mark, Fasheilach, Mount Keen, Cock Cairn, Hill of
Cat, and Mudlee Bracks. On Mount Battock the two
counties are met by a third—Kincardine. Thence,
passing Sturdy Hill, the county march descends to the
North Esk at The Burn, and follows the course of that
river to the sea.

Southward to Buddon Ness, the county is bounded by
the North Sea ; and then westward to Invergowrie, by
the estuary of the Tay.

4. Surface and General Features.

The surface of Forfarshire is highly diversified : it
contains within its bounds districts that are highland,
others that are lowland, and yet others that are maritime.
In the north extend in a north-easterly direction the
Grampian mountains, which rise ridge upon ridge from
the Braes of Angus, their southern spurs, until they attain
on the county march their highest general elevation of
more than 3000 feet. With a breadth varying from
9 to 15 miles, they stretch for 24 miles through our
county, where they are sometimes known as the Benchinnin mountains. A parallel expanse of lowland
country to the south-east, 32 miles by 4 to 6 miles, is
the fertile Howe of Angus, a portion of Strathmore.
There follows, as we advance southwards, another parallel

section, the Sidlaws, with their eastern and north-eastern spurs, a range from 3 to 6 miles broad and 22 miles long. The maritime district succeeds this, curving round the southern and eastern parts of Angus in a belt 37 miles by 3 to 8 miles.

The highland district is the largest and most picturesque. It occupies at least two-fifths of the whole

Glen Clova

shire. Through it various glens run from north to south, the chief being Glen Isla, Glen Prosen, Glen Clova, and Glen Esk. At the heads of these valleys rise the highest summits of the county, while their dividing ridges attain a scarcely lower altitude. The giant Glas Maol (3502 ft.) towers above the infant waters of the Isla. A particularly shapely, though less ambitious, hill, Mount Blair (2441 ft.), commands one of the most extensive and varied prospects in Scotland, and buttresses Glen Isla on the west. Bannoch, Broad Cairn, and others on this part of the county march look down upon the sullen waters of Dubh Loch, a deep mountain tarn in Aberdeenshire directly under " dark Lochnagar."

Forter Bridge and Mount Blair

The Prosen finds its source in Mayar (3140 ft.); and Driesh and other heights separate its glen from that of Clova. Cat Law (2214 ft.), to the south-west of Prosen and Clova, has a wide prospect extending in clear weather almost from Stirling to Stonehaven.

Next come two minor glens, Glen Ogil with the Noran and Glen Lethnot with the Water of Saughs, lower down called West Water. To the north the broad expanse of the Wirren (2220 ft.) shuts in the valley of the North Esk on the south and gives it its final eastern trend. This is a fine hill. Its ample green slopes, indented with many a mountain stream but un-scarred by these, seem to breathe peace.

On the north and east of Glen Esk there curves a noble series of rounded heights, almost matchless in the county for rugged grandeur. The Mark and the Tarf, two tributaries of the North Esk, each with its own glen, descend from the frontier heights. Conspicuous amongst these are Mount Keen (3077 ft.), Hill of Cat, and Mount Battock (2555 ft.), on whose summit three counties converge.

Forfarshire can scarcely vie in its highland scenery with western and northern Scotland. In general shape its mountains are oblong, rounded, or flat-topped. These tame braes and knocks and meals and shanks usually command vast expanses of heath-clad or barren plateaus, above which in unimpressive bosses project their loftier summits, or a sea of mountain billows severed by charac-terless depressions. But to this there are exceptions. If on a smaller scale, there are precipices and peaks that recall those of Glencoe, and corries and tarns as desolate as any in the Cairngorms. Such are Caenlochan, a sublime corry that gashes the north-eastern shoulder of Glas Maol; Craig Rennet, that soars like an Alpine precipice 1000 feet sheer above the White Water of Glen Doll; and the serrated cliffs of Craig Maskeldie overhanging the tarn of Carlochy high above Loch Lee.

Strathmore, or the Howe of Angus, as the Forfarshire portion of it is called, is a fine expanse of fertile land in the heart of the county. It is not flat, but diversified by numerous eminences. Many of these are isolated, while others extend in more or less continuous mimic ranges. On the whole they belong either to the Braes of Angus or to the Sidlaws. Amongst the former are the Hills of Kingoldrum and Kirriemuir, and the White and the Brown Caterthun; amongst the latter, the Hills of Kinnettles, Finhaven, Turin, Balmashanner, and Dunnichen. Some are wooded, and within their bounds flow tiny streams through lateral valleys. The strath contains a series of small lochs which add much to the attractions of the landscape. No finer view of the Howe of Angus can be obtained than that from a point about two miles to the south-west of Forfar on the Dundee road. The peaks of the distant central highlands of Perthshire—Ben Lawers, Schiehallion, Ben-y-Gloe, and the entire Benchinnin range from Glas Maol to Mount Battock, stretch in clearly defined outline along the west and north; the gentler heights of the Sidlaws extend in finely marshalled array to the south-west; and the wide strath with its fields and lakelets, its hills and woods, its villages and towns, is unrolled in the foreground and to right and left.

The next district takes its character and its name from the Sidlaw Hills. Starting from Kinnoull Hill, these hills enter Angus at Gask. Many of their most picturesque, though not their loftiest summits, are already passed before the bold cliffs of Lundie, whose surrounding woods are jewelled with lochs, the tower-crowned

Kilpurnie, and the wooded hill of Auchterhouse (1399 ft.) are reached, and the greatest height of the entire chain is attained in Craigowl (1493 ft.). Seen from the opposite shore of Fife, this last mountain appears to dominate the whole district and forms a grand background to the city of Dundee, on which the nearer cone-shaped Law Hill

Law Hill, Dundee

looks down. Between these two there stretches from east to west the fertile district of Strathmartine.

Three fairly well marked ridges strike outwards from the neighbourhood of Craigowl : one may be traced in the succession of heights, many of them wooded, that descend to the coast near Monifieth ; the central ridge broadens out to north and south as it advances eastwards to the cliffs at Arbroath and the Red Head near Lunan

Bay; and the third and highest extends eastwards to the South Esk, whose valley it bounds on the south all the way to Montrose.

Maritime Forfarshire is that portion of the county which lies between the Sidlaws and the Tay or the ocean. From the western marches to the mouth of the North Esk, a line traversing this region and parallel to the coast is about 37 miles long. This district is widest in the west and narrows as the hills approach the sea in the east. The ground undulates and rises not infrequently into knolls and even hills, many of which are planted with fine woods. Near the sea, between Broughty Ferry and Arbroath and again between the mouths of the two Esks, there are wide expanses of downs or links. The links are valuable for grazing, and for their numerous rabbit warrens. In certain parts they have been laid out as golf courses, of which those of Monifieth, Barry, Carnoustie, Elliot, and Montrose are noteworthy.

5. Rivers and Lakes.

Though the river Tay belongs to Perthshire, one of its main affluents, the Isla, flows for at least two-thirds of its course through Angus. Forty-five miles from its junction with the Tay, far up in the heart of the eastern Grampians, the Isla finds its source in two streams, one descending from the corry of Caenlochan and the other from Glen Cannes. At first it flows due south through a narrow, tortuous glen, but soon turns south-east through

a wider valley. Further on, it plunges into the Den of
Airlie, one of the grandest ravines in this part of Scotland.
After curving to the west, the Isla winds through the
heart of Strathmore to join the main stream. Many
hill burns, as the Brighty and the Newton Burn, pour
into the Isla. Lower down it receives the Burn of
Alyth and the Melgum. The Dean from the Forfar
Loch, and the Ericht, known in one of the main branches
of its upper course as the Blackwater—a stream nearly as
long as the Isla itself and flowing through a valley parallel
to it—are tributaries that join it in the Strath.

The Reekie Linn in the Den of Airlie is as imposing
as any waterfall in the country. At this point the river,
which has for some distance been hemmed in by precipitous banks, plunges over a rock 80 feet high in a cataract
which when swollen is unbroken in its descent. The
smoke-like spray which then enshrouds it gives its name
to the fall. Scarcely less remarkable is the boiling Slug
of Auchrannie, a rocky gorge of the river about a mile
farther down the Den of Airlie. The waters of the
Melgum flowing through the Loch of Lintrathen, from
which Dundee derives its main water supply, have been
partly impounded to increase the resources of that fine
natural reservoir.

The central river of Forfarshire is the South Esk, a
stream entirely within the county limits. It rises in Cairn
Bannoch (3314 ft.), the extreme north-western point of
Angus, is joined by a small hill burn that carries off the
overflow of Loch Esk, often mistaken from its name as
the source of the river, and under the sublime heights

Reekie Linn

of Cairn Broadland, the Scorrie, and the Red Craig, receives the White Water from Glen Doll. The valley of the South Esk is known in its upper reaches as Glen Clova. For the first half of its course the river flows south-east. At Cortachy it enters a rocky channel and a richly wooded district, and after being joined by the Prosen, it sweeps eastwards into Strathmore and winds in a tranquil course to the sea, which it reaches through the wide estuary of Montrose Basin. Its main affluent on the left bank is the Noran from Glen Ogil. About six miles from Montrose Basin the river leaves the wooded policies of Brechin Castle and curves round the southern part of the ancient ecclesiastical city of Brechin. Lower down it passes the noble deer park and grounds of Kinnaird Castle.

Like the Isla and the South Esk, the North Esk is partly a highland, partly a lowland stream. For about half its course it forms the boundary of Kincardine. Its twin head-waters, the Lee and the Unich, join each other under the precipices of Craig Maskeldie, and after flowing through Loch Lee it receives the Mark on the left and the Effock on the right. The valley of the Tarf, the main highland tributary of the Esk, spreads out like an extended hand to the north, each finger representing a hill burn descending from the frontier heights of the county. At the point where the Tarf joins the river is situated the hamlet of Tarfside, the headquarters of Upper Glen Esk. Towering above it in the west is the Craig of Migvie or Hill of Rowan crowned with an imposing monument to Fox Maule, Earl of Dalhousie. After forming a huge **S**-shaped curve, the axis of which lies east and west, the Esk enters a somewhat wider part of the glen, which it follows until it plunges into a ravine similar to those at Airlie and Cortachy. This, the most romantic part of its middle course, is crossed by the main road at the Gannochy Bridge, which presents fine views both up and down the river. Thereafter the Esk passes

Gannochy Bridge

the pretty village of Edzell and continues its devious course to the sea, three miles north of Montrose. The Water of Saughs, known lower down its course as the West Water, is the longest tributary of the North Esk and joins it on the right bank a little below Edzell. The Cruick, a shorter stream flowing almost parallel to the South Esk, enters it in the grounds of Stracathro House.

Besides these larger rivers, four smaller streams fall to be noticed. The Lunan, with its tributary, the Vinney, rises in a chain of lochs east of Forfar and has a short course to Lunan Bay. The Brothock, a still smaller rivulet, enters the sea at Arbroath, a contracted form of Aberbrothock. A little to the west, the Elliot passes by Carmyllie and Arbirlot. The Dighty has a much longer course. It rises in the Lochs of Lundie under the central Sidlaws, flows east through Strathmartine and enters the lower estuary of the Tay at Monifieth.

The lochs of Forfarshire are less numerous and less extensive than they were, say, a hundred years ago. Within that time many have been drained, with the double advantage of increasing the farming land and improving the climate of the county. In the case of lochs into which streams flow, silt is always being deposited, which through time decreases the area of the sheet of water. On the other hand, a loch is sometimes artificially enlarged and deepened for the purpose of supplying the wants of a town. Of this a notable instance is Lintrathen Loch.

There remain a few small lochs in the Lundie district. One by the roadside at Tallybaccart is the haunt of wild fowl and gulls. One or two pretty lochs lie deep in the neighbouring woods and at the head of Strathmartine is the small circular Pitlyel Loch in the midst of charming scenery. The Long Loch of Lundie directly under the highest parts of the Crags has been leased by the bleachers and millers of the

Dighty, and at certain points its banks have been raised so as to secure at all seasons a sufficient·supply of water for commercial purposes.

The Loch of Forfar, one mile and a quarter long by a quarter of a mile wide, lies close to the county town. Its surplus waters have been led into the Dean and so to the Isla. St Margaret's Inch, an island in this loch, was

Loch Lee

the site of a religious house founded by Alexander II. To the east of Forfar stretches a chain of small lakes— Fithie, Rescobie, and Balgavies—which are drained by the Lunan river.

The finest lochs are in the highland portion of the county. Every glen has its tarn. Drumore is under the heights of Mount Blair in the extreme west. In

Glen Isla there is Auchintaple. Loch Esk lies high amongst the corries at the head of Glen Clova, and on the east side of the same glen, not far from the Kirkton but 1300 feet above it, are two deep mountain lochs, Brandy and Wharral, lying each under an amphitheatre of sterile and precipitous rocks. To the east of the Wharral ridge is Stony Loch, the source of the Water of Saughs. In Glen Esk there are two small lakes bearing the name of Carlochy, one in Glen Mark and the other in a deep embrasure of Craig Maskeldie. Loch Lee is, however, the chief mountain lake of Angus. A mile and a quarter in length by half a mile in breadth, it is in its noble surroundings a typically highland loch. Heather clad or bare rocky mountains rise to a great height on its two sides and its upper end, and except where the Lee is gradually silting it up on entering its waters, it is of great depth. Like so many more of the mountain lochs of the district, it is stocked with fine trout.

In 1870 the Water Commissioners of Dundee were empowered by Act of Parliament to purchase Lintrathen Loch as a source for supplying water to their city.

Where bleach works and mills have not contaminated the purity of their waters, the lakes and streams of Forfarshire afford excellent trout fishing, which is particularly fine in the highland section of the county.

6. Geology and Soil.

Geology is the science that deals with the solid crust of the earth; in other words, with the rocks. By rocks, however, the geologist means loose sand and soft clay as well as the hardest granite. Rocks are divided into two great classes—igneous and sedimentary. Igneous rocks have resulted from the cooling and solidifying of molten matter, whether gushing forth as lava from a volcano, or, like granite, forced into and between rocks. Sometimes pre-existing rocks waste away under the influence of natural agents as frost and rain. When the waste is carried by running water and deposited in a lake or the sea in the form of muddy sediment, one kind of sedimentary rock may be formed—often termed aqueous. Other sedimentary rocks are accumulations of blown sand : others are, like stalactites, of chemical origin : others, as coal and coral, originate in the decay of vegetable or animal life. Though, in their origin, sedimentary rocks occur in layers or strata, they may receive disturbing shocks and be tilted up at various angles or even bent into folds. Again, heat, or pressure, or both combined, may so transform rocks that their original character is completely lost. Such rocks, of which marble is an example, are called metamorphic.

The general contour of Forfarshire is an index to its geology. The trend of its rocks is from south-west to north-east. In the north we have in the Braes of Angus the oldest formation, the Silurian rocks consisting of mica schist, gneiss, hardened grit, clay slate, etc. On the borders of Aberdeenshire at the heads of the valleys the Silurian strata are pierced by granitic masses. The line of the Grampians is marked by a great "fault" which extends from the Clyde to Stonehaven.

These older rocks are succeeded as we cross the county to the south-east by the Lower Old Red Sandstone formation, which forms the basis of the whole of the Lowland region, and consists of conglomerates, sandstone, marly shale, cornstone and limestone, flagstones, tilestones, and shales. Erosion has been at work everywhere, but the rocks which have had most power to resist its ravages, form such hills and ridges as the Sidlaws, Turin Hill, and other heights that run eastwards to Brechin and Montrose. The dip of the rocks on the Braes of Angus on the one hand and that of the northern side of the Sidlaws on the other, show that Strathmore in the language of the geologist lies in a great synclinal curve or trough, which is succeeded by the anticlinal curve or saddleback of the Sidlaws. The axis of the one curve runs up Strathmore from Alyth to Stracathro, while the other does not follow the crest of the Sidlaws, but may be traced from Montrose to Friockheim, Letham, and Tealing, and thence into Perthshire by way of the hills behind Inchture.

The Sidlaws, which are geologically a continuation of the Ochils, are volcanic in origin, and consequently sheets of lavas and ashes have been interbedded with the sedimentary strata of the Old Red Sandstone. Their escarpments are tough igneous rock. In this southern district of the county there has evidently been at sundry periods a very considerable amount of volcanic activity, the chief foci being at Tealing, the Law Hill and Balgay Hill (Dundee), Rossie Hill, and other heights. Innumerable veins and irregular dykes and sheets of igneous rock have burst through the Old Red Sandstone. Hence we find porphyrite at Ninewells, bedded trap-rocks at Kinpurney (Kilpurnie), Charleston, Broughty Ferry, the Laws, Panmure Hill, Arbirlot, etc., and great sheets of porphyrite stretching from Letham to Montrose. Many of these igneous rocks are tufaceous in texture and not durable when exposed, but other samples, as at Craigie, are of a more basaltic character and make excellent road metal.

It has been surmised that a wide inland sea, which has been named "Lake Caledonia," once washed the base of the Grampians; that coarser gravels and shingles brought down by streams from the Highlands were deposited near its northern edge; and that finer sediments formed closely-grained sandstones, shales, and flagstones in the deeper parts farther south. These lower series of sandstones are greyish blue and brown with shades of purple. Through time the "lake" contracted, and the salts of iron to which the colouring of the stone is due became relatively more abundant. Thus a second series of rocks was deposited, which took a deeper hue of red.

It is in the lower series, and particularly in a thin band of shale some three feet thick which may be traced to the south of the Sidlaws from Balruddery Den to Tealing and from Duntrune by Carmyllie to Leysmill, that fossil remains of fish have been so abundantly found. These also occur in a similar deposit that runs from Turin Hill through Farnell into the Mearns. The fossil fauna, which is much more abundant and interesting than the flora, comprises both crustaceans and fishes. There are three genera of crustaceans found—Pterygotus, Stylonurus, and Eurypterus—while the fishes represented in these rocks, the Cephalaspidae and Acanthodidae, kinds now extinct, have the tail fin principally developed on the under side, as in sharks. These fossils have rendered Forfarshire an important field of research to the palaeontologist. In Pleistocene clays at Carcary, Drylees near Montrose, and Barry, there occur the only other fossil deposits of the county.

The red sandy marls of the Tannadice district form the highest beds in the county and these are succeeded east and west by deposits of clay and lime that sometimes contain shells.

Forfarshire appears to have been covered during the Glacial Period by a vast ice-sheet perhaps 1500 feet thick that moved down from the Highland glens, crossed Strathmore in a south-easterly direction, surmounted the Sidlaws, which deflected it still more towards the east, and descended to the sea. Its course is marked by striae or scratchings on rocks and stones. Two important results followed. Boulder clay accumulated under the ice and remained to form the soil. Vast deposits of sand and gravel kames, as for instance between Lindertis and Glamis, were left in its wake, and broadly scattered throughout the southern part of the county from the seashore to the top of such heights as Lundie (1000 ft.) and Craigowl (1500 ft.)

are boulders of Silurian and granitic rock carried down by the ice and left, when the ice melted, at spots remote from their place of origin. When the general sheet of ice had disappeared, there must for ages have been glaciers in the Forfarshire glens. An interesting relic of these is seen at Glenairn (South Esk), where a terminal moraine 200 feet high and half a mile broad runs across the valley. Above this there must, to judge by the deposits, have been a lake; but in course of time the river burst through the barrier and drained the accumulated waters into Strathmore.

It used to be supposed that the glens of the county had been formed by dislocations of the earth's crust, but it is now believed that they have been carved out by the natural agencies of running water and frost.

The soils of Forfarshire are either primary or secondary. The first is produced by the disintegration of native rocks, and the second by the materials brought from a distance by ice or by running water. The colour varies from red to brown and black. In upland districts and on gravelly bottoms the soil is thin, while the sandstone rocks have a covering of tenacious clay. Trap rock soil is friable and fertile. Secondary soils are sometimes too sandy, at other times too stiff. Often primary and secondary soils are mixed. Accumulations of water in hollows produce mosses and bogs. If not naturally very fertile, the soil of the county by such farming operations as draining and manuring has been rendered as productive as any in Scotland.

7. Natural History.

The flora of Britain has been divided into four classes, each adapted to special climatic conditions: (1) alpine ; (2) sub-alpine ; (3) lowland ; (4) maritime.

When in far off days the ice of the Ice Age was disappearing from Europe, Britain seems twice to have been connected with the Continent by a land bridge. Across this the vegetation of the Continent followed the retreating ice. First came alpine forms, then sub-alpine, next lowland, and finally maritime. The first two classes readily obtained a footing in Britain, but while the two last were still crossing the land bridge became submerged. At the next upheaval these forms effected a crossing, and a severe struggle for existence took place between them and their predecessors. The result was that the alpine and sub-alpine species were forced into higher altitudes. Before every species had crossed, the land bridge was again submerged. The general trend of the advancing species was towards the north-west. Consequently we find that the southern and eastern portions of Britain have a greater variety of species than the north and west ; and for the same reason Ireland is poorer in species than Britain.

In Forfarshire all the four classes are well represented, and few counties, if any, surpass it in the richness and variety of its flora. This is chiefly owing to its great diversity both of elevation and of soil. In the long coast-line many variations in soil conditions are found—the muddy estuaries of the Tay and the South Esk, the stony beach, sand dunes and links at Barry, Carnoustie, Elliot and Montrose, and the cliffs from Arbroath to Montrose. Behind this and in front of the Sidlaws we have a lowland region, which is largely defaced by cultivation or laid out in towns and hamlets, but there are still some patches of natural vegetation in such places as the Den of Mains and the Birkhill Feus. The Sidlaw range furnishes admirable conditions for the growth of sub-alpine varieties. Behind this again is the long stretch of Strathmore, a second lowland region also largely cultivated, but intersected with wooded dens, and traversed by numerous streams whose sylvan banks provide special local soil conditions adapted to the growth of plants requiring shade and moisture. Behind Strathmore the ground rises gradually until it reaches an altitude of over 2000 feet—sufficiently high for alpine plants to grow.

Many very beautiful specimens of seaweeds or algae may be collected at low tide along the coast, especially on the beach at the foot of the cliffs. The transition from salt water to fresh water mud-plants may be studied at Invergowrie. The formation of firm soil from loose sand by the binding action of the roots and underground stems of the sea couch grass (*Agropyron junceum*) and the sea sedge (*Carex arenaria*) is in evidence at Barry. An interesting feature in plant life which is abundantly illustrated in Forfarshire is the similarity in structure between the maritime cliff plants and the mountain plants, several species being found growing on the cliffs and again in the mountain solitudes, but nowhere between.

The influence of man has come to be regarded as one of the conditions affecting plant associations, and is classed along with climate and soil. The cultivated land of the lowland area is divided into two regions—an upper and a lower—according as wheat is grown or not. The lower region is the region of wheat. The area of Strathmore presents a great variety of woodland, partly cultivated, partly natural. In the estates of Cortachy and Glamis, for example, we get specimens of cultivated woodland consisting for the most part of deciduous trees. Oakwoods clothe the river banks in the lower regions, and higher up the shimmering birches enhance the valleys with their sylvan beauty, as for example in Glen Prosen, Glen Clova, and Glen Esk. Large tracts of Scots pine are found scattered over the county, notably at Montreathmont Moor—remains of the extensive forests that formerly clothed Forfarshire. At considerable elevations on the moorland the larch is fairly conspicuous. The beech, the sycamore, the lime, and the horse and Spanish chestnuts are common, but are not indigenous as are the oak, the ash, the elm, the rowan, the hazel, and the alder.

The vegetation of the Sidlaws consists of sub-alpine plants; heather being dominant on the basalt and sandstone soils, for example, on Craigowl and Auchterhouse Hill, and grasses on andesitic soils, where whin and bracken are also plentiful. Amongst the heather areas there is a marked absence, as a rule, of peat-bogs; and on the grass-covered hills it has been noticed that the grasses are coarser and stronger on the north side of the hill than on the southern.

On the Esk

It is, however, the mountain fastnesses, and high table-lands on the northern and north-western borders of the county that are of special importance to the botanist. There we find typical alpine vegetation, with its low growth, small leaves of a fleshy or hairy nature, and wiry stems. The commonest specimens are ordinary ling, various heaths, blaeberry, cowberry, crowberry. Another mountain-dweller is *Rubus chamæmorus*, the cloudberry or avron. The alpine lady's mantle (*Alchemilla alpina*) frequently forms a green carpet. Five different species of saxifrage, and fifty species of *Salix*, chiefly alpine, occur in the county. The peat-bogs are enlivened by brightly coloured mosses, and red glistening sundews; and occasionally the white petals of the Grass of Parnassus spread themselves out in unsullied purity above the black marsh.

The rarities of the county are the snowy gentian (*Gentiana nivalis*) which is found in the Caenlochan district, the only other recorded locality in Britain being Ben Lawers. The same district is the home of such rare ferns as the holly fern (*Polystichum lonchitis*), green spleenwort (*Asplenium viride*), and *Polypodium alpestre*. The mountain brittle bladder fern (*Cystopteris montana*) is very rare, but has been found in Glen Caenlochan. *Polypodium flexile* has been met with in Glen Prosen, and with the exception of Ben Alder this is its only known British haunt. Two rare grasses, the alpine fox-tail grass (*Alopecurus alpinus*) and the alpine cat's-tail grass (*Phleum alpinum*), are found on the banks of the Feula burn. The red alpine campion (*Lychnus alpina*) grows on Little Culrannoch, a hill northward from the head of Glen

Doll. So far as is known this plant occurs nowhere else in Scotland, but a similar plant has been met with on the crags of Helvellyn. In a ravine on the south side of Glen Doll is found the blue alpine sow thistle (*Mulgedium alpinum*)—a very rare species growing also on Caenlochan and on Lochnagar. Glen Doll yields two other rare treasures—*Oxytropis campestris*, a small vetch with no other habitat in Britain ; and the alpine milk vetch (*Astragalus alpinus*), found only here and on the Braemar hills.

Mosses and ferns are well represented in Forfarshire, indeed this county has a greater number of species of mosses than any other county in Great Britain, Perthshire alone excepted. Most of the common ferns abound ; and besides the rare forms already mentioned, the filmy fern (*Hymenophyllum Wilsoni*) is got at the Reekie Linn, and on the Caenlochan mountains, while the moonwort is abundant on the Links of Barry, and is found in various other localities. The adder's tongue (*Ophioglossum vulgatum*) is said to occur on Barry Links, but is rare. The royal fern (*Osmunda regalis*) has been found at Arbroath, Montrose and Kinnaird.

It was not only a fresh flora but also a fresh fauna that Britain received after the ice sheet disappeared ; with this important difference, however, that forms which could fly or swim were not checked by the submergence of the land.

Forfarshire possesses the common wild creatures of Britain. Rabbits and hares are plentiful. In winter and early spring the white hare is a common sight on the mountains and moors. These mountain hares belong to a different species from the lowland hares. This species, which turns more or less white in winter, is found throughout the palaearctic region along with arctic plants, and its presence, often in isolated positions, is regarded as one of the proofs of a former glacial epoch.

Another inhabitant of the mountains and moorlands is the red deer. Large herds of these noble and beautiful animals roam over the hills, but unless taken by surprise it is very difficult to see them at close quarters. Roedeer are native but fallow deer have been introduced. Badgers are gradually dying out. Stoats are abundant. Weasels, moles, hedgehogs, and squirrels are common. Mice and voles and the grey rat are so abundant as to have become a pest in some places. The grey rat has almost extirpated the black, a few specimens of which, however, yet remain.

On account of the variety of environment, there is scope for mountain birds, for sea birds and for cliff birds, while the dens and woods of the rural districts are alive with countless varieties of our native songsters—blackbirds, thrushes, finches, linnets, buntings, wrens, larks, titmice, stonechats, whinchats, pipits, etc. The long-eared owl, the brown, the white, and probably the short-eared, are native, while three varieties of swallows along with the swift are summer visitors. From the verdurous gloom of the spring-woods, one often hears the clear call of the cuckoo and the cooing of the wood-pigeon. Rooks and jackdaws are very numerous at all seasons, and even the magpie has been seen, although rarely. Corncrakes, partridges and pheasants are common in the fields. The capercailzie is found in some places, and on high hills the ptarmigan. The curlew, plover, grouse and snipe are common. Higher on the mountains bird preys on bird ; the eagle barks from his lonely eyrie, the hawk circles above his intended prey, and the sparrow-hawk pursues his victim to death. Besides these we may see the buzzard, the kestrel, and occasionally the peregrine falcon.

Aquatic birds are numerous at the mouth of the Tay, on the links, sands, and cliffs. Many of them are winter migrants, such as the scaup, pintail, and widgeon, the blackheaded gull, and others. The tern or sea swallow, a beautiful bird which skims over the surface of the water with very rapid flight, is a summer migrant, whilst the arctic tern comes in autumn. Along the cliffs and on the rocks at sea the gannet or solan goose appears in spring and autumn. Vast colonies of gulls—common, blackbacked, kittiwake and skua—razorbills, puffins, guille-mots and ducks of diverse kinds make the whole coast-line resonant with bird life, while the storm-petrel warns the fisherman of a coming tempest. The gulls do not confine themselves to the coast but fly far inland ; and the black-headed gull breeds on inland lakes and marshes. The hooded and carrion crow are more common near the coast than inland.

The marine fauna of Angus is that of the east coast in general. Seals are plentiful in the Tay estuary and can be seen at low tide basking in the sunshine on the sand banks. Whales also have repeatedly visited the Tay, and on more than one occasion have been attacked in the river at Dundee. The well-known Dundee whale was of the humpbacked species (*Megaptera longimana*). It was pursued and wounded at the mouth of the Tay, and was subsequently found dead off the coast of Kincardine. Its skeleton is preserved in the museum in Dundee.

8. The Coast Line.

The coast-line of Forfarshire, about 37 miles in length, presents great variety of feature—alluvial shores, sands, cliffs and raised beaches.

The Carse of Gowrie, for example, at the eastern extremity of which we may begin our peregrination, is the levelled terrace of the 50 feet raised beach, now turned into rich alluvial land. At Dundee the lower parts of the town are built on the 25 feet beach. On platforms of the same height rest the blown sands of Barry, Carnoustie, and Montrose. The 100 feet beach can be well seen at Barry, Arbroath, and Montrose.

Between Dundee and Broughty Ferry, interesting primeval deposits have been unearthed at Stannergate in the form of "kitchen-middens"—the refuse heap of some primitive community. At Broughty Ferry a small rocky promontory juts into the firth and narrows it to about the breadth of a mile.

Below Broughty Ferry the Lady Bank runs out to a sharp spit of sand and forms one side of the small estuary of the Dighty. On the landward side the Sands of Barry, with two lighthouses, stretch out to Buddon Ness and doubling it extend to Carnoustie. At Buddon the great sand dunes attain an altitude of 95 feet and form the most conspicuous objects on the coast-line. Between the shore-line and the railway is a great expanse of links utilised by the War Office for annual encampments and artillery practice. Close at hand are the golf courses of Monifieth, Barry, and Carnoustie. Farther east Westhaven and Easthaven, almost contiguous with each other

Fisherman's Cottage, Westhaven

and with Carnoustie, are picturesque fishing villages. Their shore is characterised by shelving rocks, which again give place to links and sands as Elliot with its golf course and Arbroath are approached.

About a mile beyond the harbour of Arbroath we reach Arbroath Ness, the beginning of the cliffs which render this in many ways the most interesting part of the coast. At the Ness is St Ninian's holy well, a favourite pilgrim resort in former times. Close by is the site of St Ninian's—locally St Ringan's—Chapel. In 1842 a wonderful stalactite cave was accidentally discovered in the Ness Quarry. Farther east is the Needle's Eye, a curiously perforated rock ; and a great ravine, called the

Dickman's Den

Cruzie from its resemblance to an old Scottish lamp. At the Blow Hole the sea waves rise in storms to the height of 150 feet. The Smuggler's Cave, and Dickman's or Dickmont's Den, were in the eighteenth century the haunts of smugglers. The Three Storied House and the Mariners' Grave are caves with names that tell their own tales. A great shore stack, separated from the adjoining

cliff by a passage called Duncan's Door, and one of the
most remarkable pieces of rock scenery on this coast, is
variously known as the Deil's Head and the Pint Stoup.
The Masons' Cave, 231 feet by 12, was long a place
of meeting for the St Thomas Lodge of Freemasons.
Another is suggestively called The Forbidden Cave, into
which, according to tradition, a piper and his wife,

Red Head

regardless of the prejudice against entering its precincts,
wandered never to return. Perhaps the most awe-inspir-
ing of the Arbroath caves is the Gaylet Pot. This huge
cavity, about 100 yards from the sea and in the midst of
an arable field, communicates with the ocean by a tunnel
130 feet below the summit of the cliff. When in storm
the sea rushes through this subterranean passage and boils

in the bottom of the crater-like chasm, it strikes such horror into the beholder as he will long find difficult to rid himself of.

Half a mile beyond is the quaint village of Auchmithie, which disputes with the neighbouring Ethie Haven the distinction of being the "Musselcraig" of Scott's *Antiquary*, as does the adjacent Newbarns with Hospitalfield for being the original of the "Monkbarns" in the same novel. Ethie House, near Auchmithie, is by common consent the residence of "Sir Arthur Wardour." The Red Head, 267 feet high, the most imposing sea cliff in the county, terminates this rocky section of the coast.

We now reach the fine curve of Lunan Bay, the beautiful sands of which stretch for some five miles from the Red Head to Boddin Point, where cliffs again occur. Between this and Scurdie Ness at the mouth of the South Esk there is to be seen on the shore the huge mass of the rock of St Skeoch, called also the Elephant Rock from its striking resemblance to that animal. High above on the cliff is that most romantic of burial-places, the little churchyard of St Skeoch. The shore of Usan Bay is strewn with rugged masses of rock. Round Scurdie Ness and within the estuary of the South Esk is the prosperous fishing village of Ferryden, opposite Montrose. Beyond the river a long line of fine sands, flanked by the famous golf links of Montrose, extends to the mouth of the North Esk, where the coast-line of the county terminates.

9. Coastal Gains and Losses. Sand= banks. Lighthouses.

The Forfarshire coast affords striking examples of both accretion and erosion. The gain has been partly due to natural causes, but in the case particularly of the foreshore at Dundee a large area has been reclaimed. The accompanying plan of the river frontage at Dundee shows how much has been accomplished since 1793. Not only the railways, docks, and shipbuilding yards, but also much of the lower area of the city is situated on entirely made-up ground. Thoroughfares with names like Yeaman Shore and Seagate, now at a distance from the river, once flanked its waters, and boats were drawn up at a pier still in existence but now far from the water's edge. A long sea-wall extending practically from Ninewells to Stannergate, a distance of some five miles, presents a front to the river, broken only by the entrances to the harbour. The esplanade or western part of this structure is a broad and handsome promenade. At present the harbour authorities have under consideration a further addition which will win from the river a great acreage of dock accommodation. The tidal lands reclaimed by the Harbour Trustees since 1830 are 188 acres in extent ; besides which the original owners of the foreshore made up 61 acres, and the Town Council 124 acres. Large as are these areas, their extent is small when compared with what might yet be done ; for it is estimated that in the Firth of Tay some ten or twelve square miles of land are laid bare at low water.

In the neighbourhood of Monifieth and Buddon Ness a considerable amount of erosion takes place. The shore is exposed to violent south-easterly gales, and as recently as February, 1912, a great deal of damage was done to the promenade along the river side. The lighthouses at Barry were formerly erected at the southern extremity of Buddon Ness, but on account of the shifting and wasting of the shore in the neighbourhood they have been removed about a mile and a quarter farther north. The spot on which the outer lighthouse stood in the beginning of the seventeenth century was found in 1816 to be two or three fathoms under water and quite three-quarters of a mile within flood mark. At this part of the shore, however, the sea gives up again much of what it carries away. Recent measurements put the erosion at 56 acres and the accretion at 46. To check this loss groynes have been erected by the War Office in the vicinity of Buddon, and near Broughty Ferry by the local municipal authorities.

Between 1786 and 1816 the shore road to the west of Arbroath had twice to be shifted owing to erosion of the coast; while immediately to the south-west of the town the foreshore in the beginning of last century was being encroached on to the extent of about one yard per annum.

In the north of the county the most extensive area that lends itself to reclamation is within Montrose Basin. In 1760 a scheme was projected for reclaiming 2000 acres. A dyke called the Drainer's or Dronner's Dyke was built across the basin, which being promptly destroyed by a storm the idea was abandoned.

The estuary of the Tay and the coast of Forfarshire, fraught with much danger to mariners, have 22 charted lights. The Firth of Tay contains so many sand-banks that its navigation is difficult. Fortunately for Dundee these hazards occur chiefly in the upper and broader part of the estuary. The channel which must be kept to by river steamers and the craft that sail up to the harbour of Perth is near the middle of the river. Lights on the Tay Bridge mark it at night.

Lights on the ferry piers at Dundee and Newport, on a floating buoy that marks the end of a dangerous bank in mid-river, and on the Beacon Rock near the north side, are under the control of the Dundee Harbour Trustees. Outside of the narrows of the estuary between Broughty Ferry and Tayport are the two lights of Buddon Ness, each of 7000 candle power. The High Light, 103 feet above high-water mark, can in clear weather be descried at a distance of 16 miles; and the Lower Light, 61 feet high, at a distance of 13 miles.

Off Abertay Spit a lightship is anchored in $5\frac{1}{2}$ fathoms of water. It is coloured red, has two masts, and the word *Abertay* painted on its sides. In fogs a siren on board emits three piercing blasts every three minutes.

One of the most dangerous reefs on the Scottish coast is the Inch Cape Rock, 13 miles east of Buddon Ness and about 12 south of Arbroath. It is some 1500 feet in length and 300 in breadth, and its highest part, covered at spring tides to a depth of 12 feet, protrudes 4 feet above water at ordinary times. In old days mariners were warned of its dangers by a bell tolled by the action of the waves, and hence the name "Bell Rock." According to tradition this bell, which had been placed there by one of the Abbots of Arbroath, was wilfully cut off by Ralph the Rover, a noted pirate, who himself, as Southey's ballad tells us, perished on the reef he had intended to be a death-trap to others. In 1807 the Lighthouse Commissioners constructed a lighthouse on the Bell Rock.

The Bell Rock Lighthouse

The white tower of the lighthouse is 117 feet high and in clear weather is visible at a distance of 15 miles. At night an alternating red and white light flashes over the waves every thirty seconds. This light, with a 392,000 candle power, is one of the most brilliant on our coasts.

The entrances to the harbours of Arbroath and

Montrose are each marked by four lights. Near
Montrose is Scurdie Ness Lighthouse, 124 feet high,
with a fog bell and a white occulting light of 13,000
candle power, visible from a distance of 17 miles.

10. Climate and Rainfall.

By the climate of a district is meant its average
weather. This is largely determined by its temperature
and its rainfall, and very much depends on its altitude and
its distance from, or proximity to, the sea. No British
county is large enough to have a distinctive climate of its
own : each shares in that of the geographical region to
which it belongs.

The following table, compiled from the late Dr Buchan's
papers, shows the mean temperature over forty years ending
1895 of several stations in Forfarshire ; and, when allow-
ance has been made for local conditions, it illustrates how
uniformly temperature varies with altitude.

Station	Height above sea-level	Temperature
Montrose	14 feet	46·8° F.
Barry	38 ,,	46·8° ,,
Arbroath	68 ,,	46·5° ,,
Dundee	164 ,,	46·4° ,,
Sunnyside Asylum, Montrose	200 ,,	46·2° ,,
Kettins	228 ,,	46·2° ,,
Lednathie	720 ,,	44·1° ,,

The monthly details contained in the same papers prove that in Forfarshire July (57° F.) is the warmest, and January (37° F.) the coldest month, while the mean annual temperature is 46° F. They indicate also the marked differences of temperature at various periods of the year ; for example, after the equinoxes, or between May and June. Dundee is here chosen—it is typical of the others—and the monthly mean over the same forty years is given in degrees Fahrenheit.

Dundee

Jan.	Feb.	Mar.	Ap.	May	June	July	Aug.	Sep.	Oct.	Nov.	Dec.
36·5	37·7	39·5	44·4	49·3	55·8	58·5	57·4	53·5	46·5	40·3	37·2

The cultivation of a country does a great deal to modify its climate. The draining of bogland and the destruction of forest have made Forfarshire a much drier region. On the other hand, sheltering plantations have been fostered, which protect the land from the biting east winds so common in the spring and early summer months. Fogs and hoar frosts are not infrequent, and the latter do considerable damage to the root crops. Most of the lowland part of Forfarshire has a southern slope, a circumstance that contributes to the geniality of its climate.

The most recent rainfall map, published by the Scottish Meteorological Society in 1911, is here reproduced. It embodies investigations that have been made at 129 stations throughout the country during the forty years from 1871 to 1910, and at a very much larger number of stations for shorter periods within that time.

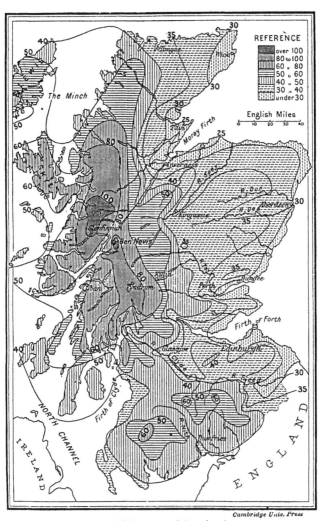

Rainfall Map of Scotland

(*By Andrew Watt, M.A.*)

The isohyets, or lines separating districts that have approximately equal rainfall, divide Scotland into parts variously shaded according as their mean annual rainfall increases from 25 to over 100 inches. Study of the map shows that on the whole the rainfall of Forfarshire ranges from 30 to 35 inches per annum, and that this average is greatly exceeded as one proceeds westwards.

A few statistics regarding the rainfall of Forfarshire, more minute than can be shown on a general map of the country, may here be given.

Stations	Height above sea	No. of years	Mean
Montrose, Sunnyside Asylum	200 ft.	40	28·89
„ Lighthouse	124	40	24·72
Dundee Waterworks			
Craigton	484	40	37·56
„ Hillhead	570	40	36·07
Crombie Reservoir	522	40	37·34
Dundee, Eastern Necropolis	199	40	28·93
Dundee Waterworks			
Lintrathen	700	30	34·63
Creich Hill	1482	30	32·58
Broughty Ferry	...	30	27·04
Lednathie, by Kirriemuir	720	20	43·17
Arbroath, Dishland Hill	69	20	23·39
Buddon Ness	17	20	25·31
Dundee Harbour	14	20	27·54
„ Dens Works	119	20	27·53

From this table it appears that the average rainfall in Forfarshire is 30·29 inches, the highest mean being 43·17 and the lowest 23·39.   Between 1891 and 1900 the greatest average rainfall for Scotland was recorded at the Ben Nevis Observatory, viz. 168·98.   From 1871 to 1880 the lowest average was at Cromarty, viz. 23·18. Maximum and minimum quantities for any given year are also interesting.   The maximum record for any year since 1871 is 240·13 at Ben Nevis Observatory (1898); and the minimum 14·72 at Pentland Skerries Lighthouse (1895).   The corresponding figures for Forfarshire are : maximum, 54·15 at Craigton Water Works (1872); minimum, 15·51 (1887) at Montrose Lighthouse.   The year 1911, which the man in the street may have thought to be abnormally dry, the more exact meteorologist has pronounced to be very ordinary in respect of rainfall. But in that case the rain must have been most unevenly distributed, for in that year the Dundee district—i.e. southern Forfarshire and the adjacent parts of Fife— received the smallest quantity of rain in the United Kingdom, viz. 17·08 inches.   The general conclusion then is that our county is one of the least rainy parts of Scotland.

The great storm of December 1879 that wrecked the first Tay Bridge gave Dundee an unenviable notoriety, and though almost equally violent winds have since occurred their effects have fortunately been less disastrous. Wind records kept at Dundee for the decade ending in 1910 show that the prevailing wind is from the south-west, the same quarter as sent us these tremendous visitants.   Twenty-eight per cent. of the registered winds have come from that direction ; 19 per cent. from the south-east ; 15 from the north-west ; 14 from the west ; and 13 from the north-east, one of the most biting experienced ; while only 1·9 per cent. blow from the north and 1·2 from the south. South-easters predominate in April, May, and June ; when it frequently happens that a warm day with westerly breezes is succeeded towards evening by a cold blast that rolls up dense, chilly fogs—"the east haar"—which envelop the eastern and southern parts of the county and penetrate to its mountain boundary. For the remaining nine months the south-west breezes attain the highest records.

The sun is above the horizon in Britain for about 4400 hours in a year, and of this possible number of hours of sunshine Forfarshire has enjoyed for the last ten years an average of 1374·4 or more than 31 per cent., which if distributed equally over the year would allow $3\frac{3}{4}$ hours to each day. The highest record is 1480 in 1905 and the lowest 1236 in 1902. Monthly averages range from 184·1 hours in June to 40·6 in December. May, June, and July are the brightest months ; and November, December, and January the dreariest—a matter of common observation which scientific research corroborates.

11. People—Race, Dialect, Population.

We possess no certain knowledge about the earliest inhabitants of Forfarshire. It lies in what was in Roman times the land of the Caledonians, who were also styled

Picts. What the Picts were racially is still an unsolved problem. Their language was of the Celtic type—as far as can be gathered from the scanty fragments that exist. *Pett* or *Pitt* (a Pictish name equivalent to "place," or "dwelling") is frequently met with in place-names in the north-east of Scotland, and is abundant in Forfarshire. A few specimens are—*Pittendrich, Pitarrow, Pitmuies, Petreuchie, Pitcur, Pitalpin, Pitscandly, Pitpointie*. The words *Angus, Mearns, Dunnichen, Monikie, Meigle* are Pictish; so, too, is *aber*, as in *Aberlemno, Aberbrothock, Arbirlot*; whereas the western or modern Gaelic equivalent of *aber* is *inver*, of which we have instances in *Invergowrie, Invermark, Inverquharity, Inverarity, Inverkeillor*, and many more. The names of rivers and mountains in Forfarshire are for the most part of a Celtic origin—*Esk, Isla, Dean, Tay, Mark, Tarf, Prosen*, etc.; while its mountains have Celtic names in the north (*Cairn na Glasha, Wirren, Driesh, Mayar, Caterthun*, etc.) and Old English ones in the south (*Sidlaws, Catlaw*, Dundee *Law, Oathlaw*, etc.).

From the dawn of history on to practically our own day there has been a considerable intermixture of races in our district. And here it must be stated that the limits of a single county are somewhat narrow for the subject under consideration. Except in special circumstances, it is safer to think of this modification of race as affecting all southern Pictavia,—Fife, eastern Perthshire, Forfarshire, and the Mearns,—rather than Forfarshire alone.

In the fourth century Frisian settlers from the

Continent began to appear on the east coast of Scotland. Hence in Forfarshire we have such names as *Westhaven*, *Easthaven*, *Ferryden*, and *Ethiehaven*. From those Frisians the fishing communities of the east coast from Fife Ness to Stonehaven are doubtless descended. Their speech differs in at least its intonation from that of landward districts ; and they are a people apart from others of the same region, intermarrying with one another and forming a distinct community.

At one time the Angles of Bernicia seemed likely to dominate Scotland ; but in 685 they were overthrown at the battle of Nechtan's Mere, identified with Dunnichen in Forfarshire. In the ninth century Picts and Scots united under Kenneth Macalpin, which effectually put an end to the warlike inroads of aliens into Pictavia.

The occurrence of the word *ness*, for headland, in some parts of Forfarshire—particularly in *Buddon Ness*—would seem to point, if not to occupation of the maritime district by the Danes, at least to their familiarity with the coast-line. Other Danish names—*Ravensby* near Barry and *Hedderwick*, *Buttergill*, and *Shieldhill* in a line stretching westwards from Montrose—give some colour to the tradition that the Danes entered Forfarshire and marched through it to Luncarty, where they suffered defeat. But there is little beyond this to bespeak an infusion of Viking blood in the district.

Within a century after Carham (1018) the Anglian folk of the south-east by peaceful immigration and settlement overspread the district from the Forth to the

Grampians, a fact of the utmost ethnological importance to Forfarshire, whose people henceforth became predominantly Lowland instead of Highland, Anglians instead of Celts. The civilization that we associate with Malcolm Canmore, his English Queen Margaret, and their immediate descendants, was admittedly English in its character and tendencies. Under those monarchs Norman families settled as a semi-foreign aristocracy in Forfarshire; but these families did little or nothing to modify the ethnology of the district. Far more important was a settlement of Flemings in the twelfth century. They introduced the weaving of linen and woollen goods. Evidence of their presence in Forfarshire is to be found in such place names as *Friockheim*, *Letham*, *Craigo*, and *Cruick*.

In comparatively recent times Celts from the Highlands made their way through the Grampians and settled in the adjacent districts. Such were the Ogilvys, Farquharsons, Robertsons, Stewarts, and Murrays.

While the race of the aborigines in Forfarshire is thus largely a matter of conjecture, within historical times its people have been essentially Anglo-Pictish, slightly modified by the immigration of Anglo-Normans, Flemings and Highlanders.

Anglian influence in Forfarshire is not confined to ethnology; it is even more pronounced in the speech of the people; for Lowland, or Braid Scots, the folk-speech of the county, is directly descended from northern English, or Northumbrian. This gradually spread northward from the Lothians. Different districts of Scotland have developed certain peculiarities, which have given rise to local dialects. North-west Fife, East Perth, and West Forfar belong to the North-Mid-Lowland district, while East Forfar with most of Kincardine belongs to the Southern-North-Lowland district.

In East Forfar *red heads* is pronounced *reed heeds*, and in West Forfar *raid haids*. Another distinction would appear to be an increase of lip-protrusion—rounding—in the pronunciation of *ui* spellings (especially before *r*), as we go east and north towards the mid-north or Aberdeen Lowland, in which after *k* and *g* the *ui* sound by further lip-protrusion changes into *wee*; e.g. *school, skuil, skweel.*

The following are some of the most striking features of Forfarshire speech :

(1) What looks like a singular verb is used with a plural noun : e.g. "The men *works* at so and so." Historically this is a survival of the original Northumbrian plural.

(2) Words beginning with *kn* have not yet lost the sound of *k* : e.g. the *k* in *knee* is pronounced. In some parts such a *k* has been changed to a *t* : e.g. *knock* is often *tnoc.*

(3) In such words as *butter, water,* the *t* is not pronounced at all ; instead, the breath in the throat is completely shut off, and sounds are heard like *bu'er, wa'er.* This is very common in Dundee, and seems to be absent, or at least less common, in the eastern part of Forfarshire.

(4) *At* is used for *that* (relative)—a Danish idiom.

With a population of 281,419, Forfarshire stands fifth on the 1911 list of Scottish counties. It is exceeded by Lanark, Midlothian, Renfrew, and Aberdeen, which have respectively 5·1, 1·8, 1·11, and 1·10 times as many inhabitants; while it is 37·2 times as populous as Kinross, which comes last on the list.

Large as this population is, it shows a decrease of 2663 as compared with the figures for 1901. This is the first time a decrease has been recorded : in previous census years the increase has ranged from 20,000 to 30,000 within a decade. The registrar-general notes that this falling off is in part due to the fact that the 1901 returns included a large number of workmen temporarily resident in Lintrathen in connection with the Dundee Water Works ; but more important causes are emigration and centralisation—both to some extent arising from depression in trade.

The position of Forfarshire on the census list is the result mainly of the industries and manufactures of its towns. More than 84 per cent. of its inhabitants live in its nine burghs, and 58 per cent. in Dundee, the largest. Dundee has a population of 165,006 ; and if Broughty Ferry, Monifieth, and Carnoustie be added, to say nothing of the townships on the southern bank of the Tay—and all these are largely residential suburbs of Dundee—some 20,000 must be added to the Dundee figures. In round numbers, the Dundee centre has about 200,000 inhabitants ; and on the strength of its own figures it ranks third among Scottish towns. In 1821 Dundee had a population of 30,575 ; in 1841, of 62,794 ; in 1871, of

120,724; in 1911, of 165,006. It has thus increased more than five times in ninety years.

The decrease of burghal population in Forfarshire is not here, but in the smaller towns. The increase in Dundee is 1·2 per cent., in Carnoustie 3, in Broughty

Broughty Ferry

Ferry 5·5, and in Monifieth (Parish) no less than 45·2— one of the most remarkable cases of increase in the whole country. On the other hand, the chief percentages of decrease are : Forfar 4·8, Brechin 5·8, Kirriemuir and Arbroath 7·8, and, largest of all, Montrose 11·7.

12. Agriculture — Main Cultivations, Stock, Woodlands.

Eleventh in point of area, Forfarshire occupies the same rank in the percentage of its cultivated land, 44·5 per cent. as against 24·2 for the whole country.

Within the last century and a half much has been done both in the reclamation of arable land and in the improving of such accessories of farming as draining, fencing, the making of service roads, and the erection of commodious buildings and steadings. Along the Braes of Angus and amongst the Sidlaws, extensive reclamation was carried out between 1870 and 1880; and thus the arable percentage of the county was raised from 41·8 to 44·5, an increase that represents something like 1246 acres each year.

In the following tables, which contain the returns for 1911, the four chief Scottish counties are given. From this the position of Forfarshire as a farming district will readily be seen :

	Wheat		Barley		Oats
	acres		*acres*		*acres*
Fife	12,992	Forfar	24,300	Aberdeen	192,416
Forfar	10,670	Berwick	18,693	Perth	67,452
Haddington	7,658	Fife	16,898	Banff	50,541
Perth	7,005	Aberdeen	16,551	Forfar	49,791

	Rotation Grasses, not for Hay		Potatoes
	acres		*acres*
Aberdeen	234,912	Forfar	16,992
Forfar	61,216	Fife	16,342
Perth	59,921	Perth	15,102
Wigton	58,825	Haddington	9,015
	Turnips		Horses
	acres		*no.*
Aberdeen	86,514	Aberdeen	30,390
Forfar	32,025	Perth	12,759
Perth	26,608	Forfar	10,126
Berwick	25,332	Fife	10,107

Forfarshire is thus one of the four chief Scottish counties in respect of no fewer that seven of the categories under which farming statistics are arranged by the Board of Agriculture : it stands first in two—barley and potatoes ; and second in three—wheat, rotation grasses not for hay, and turnips.

The returns of the Board of Agriculture not included in the above tables are those for rotation grasses for hay, permanent grasses, cattle, sheep, and pigs. In these Forfarshire is lower than fourth on each list, but yet in most has a high place. The county has altogether 112,709 acres under grass, 53,683 cattle, 167,450 sheep, and 8255 pigs. In the three last there is a marked increase as compared with 1910.

For its importance as a grain-producing district, Angus is indebted to the enlightened efforts of its great landowners and of various agricultural associations. Its green crops, particularly potatoes, are remarkably fine. The cultivation of potatoes is popular on account of its speculative character, prices ranging from £1. 10s. a ton in one year to £5 in the next : as one farmer put it— "I've sold potatoes at 13 shillings and at £13 a ton !" The soil of the county seems to be specially adapted to this crop. On many farms the tubers are sprouted in boxes for the early market, and a heavy trade is done in certain varieties.

In cattle Angus has had a reputation that may be said to be world wide. The polled or hornless cattle are familiarly known as Angus Doddies. They are also called Humble Cattle or Humlies. Probably the earliest notice of the cattle of Angus is in 1684, which shows that they have been carefully bred in this quarter for more than two hundred years. But it was in the nineteenth century in particular that this was carried to perfection. In the year 1865 the rinderpest worked havoc on the Forfarshire breeding farms and thereafter some of the splendid pedigree herds were finally dispersed. But by crossings and importations much improvement was again effected, and progress in breeding has on the whole kept pace with progress in agriculture. The cattle of the county to-day are well-bred crossed shorthorns ; but Aberdeen-Angus, and Herefords are rapidly

Aberdeen-Angus Bull

becoming favourites. As will be inferred, much more attention has been given to the feeding than to the milking breed in Forfarshire, a circumstance which differentiates the farming of the county from dairy districts like Ayrshire. Prior to the nineteenth century, the sheep of Angus were of the white-faced variety; but these came to be superseded by the black-faced sheep of Peeblesshire, and Border-Leicesters are now largely bred. Goats, once common, had to be exterminated owing to the damage done by them to plantations.

Market gardens and orchards, though not a distinctive feature of the Forfarshire countryside, are plentiful. In the neighbourhood of the towns there are many highly successful nursery and market gardens. On the western borders of the county, the cultivation of small fruits is rapidly increasing.

In early times Angus seems to have been a densely wooded district, but the primeval forest has disappeared. The royal forests of Angus were celebrated. They included Drimmie, Kingennie, Kilgary (Menmuir), Kingoldrum, Plater, and—largest of all—Montreathmont. Large tracts of coniferous trees, mainly Scots pine, are common in the Sidlaws and on the Braes of Angus; but the most extensive is on Montreathmont Moor to the south of Brechin. Oak trees abound in the lower parts of Glen Isla and along the skirts of the Grampians, and birches in Glens Prosen and Clova; while the splendid parks of Kinnaird, Panmure, Glamis, Gray, and many others of less area are beautifully wooded with mixed deciduous trees, particularly beech and oak. Some of the noblest individual trees are to be found on these estates. At Gray House there are three noteworthy trees, an oak with a height of 65 and a girth of $26\frac{1}{8}$ feet, an ash 110 by $18\frac{1}{4}$ feet, and a sycamore 81 by $15\frac{1}{8}$ feet.

About one-twenty-third of all Scotland is wood; of Forfarshire, one-nineteenth—i.e. 30,068 acres.

13. Industries and Manufactures.

The chief industry of the Forfarshire towns is the manufacture of linen goods, and for these fabrics the county has a world-wide reputation. If it be remembered that in respect of this manufacture amongst Scottish counties Fifeshire is second only to Forfarshire, and that these two shires are separated only by the estuary of the Tay, the great importance of the whole district as a textile centre will be at once manifest. It is doubtless because the soil of Angus was so well suited for growing flax that, in days when spinners and weavers depended for their raw material on home production, the industry should have established itself in this district. But manufacturers now look to foreign countries for the supply of the raw material. In the early days, moreover, that is in the closing decades of the eighteenth and in the opening ones of the nineteenth century, the work was largely carried on in rural districts, the burghs then serving as markets for what was spun and woven in country hamlets. At that time such parishes as Barry, Monifieth, Coupar-Angus, Dunnichen, Kirkden, Logie-Pert, Glamis,

Kinnettles, Mains, Menmuir, and Stracathro had scarcely a village where the inhabitants were not mainly weavers. The small farmer grew the flax, his wife spun it, and in winter-time, when out-door work did not demand his energies, he himself turned weaver. There was of course much spinning and weaving in the towns too; but country people brought their yarns and their woven goods to the nearest town—Dundee, Kirriemuir, Forfar, Arbroath, Brechin, or Montrose—and sold them in open market. The time came when every piece of cloth was inspected and stamped by regular officials before it could be produced in the markets, a practice which enhanced the value and the reputation of the productions.

The application of steam power and the invention of spinning and weaving machines produced great changes in Forfarshire. These together with the almost universal employment of foreign raw material practically put an end to textile work in rural districts. The click of the weaver's shuttle is no longer heard in the country cottage; such rural mills as gave employment to country weavers now stand tenantless or have long been utilised for other purposes; and, worst of all in every respect but that of the vast development of trade, urban population has enormously increased at the expense of the rural districts.

As early as 1526 Hector Boece speaks of Dundee as a town "in which the people travel very painfully about making and weaving of cloth." In 1727, when the manufactures of other Forfarshire towns were yet in their infancy, Dundee turned out 1,500,000 yards of linen. In 1738 Arbroath, previously little more than

a village, hit almost by accident on the production of a
cloth known technically as osnaburgs.   After 1746 the
flax trade rapidly prospered in Brechin, where also osnaburgs were the chief fabric made.   In the same year
Forfar began the same manufacture.   About this time
the manufacture of brown linen was introduced into
Kirriemuir, and carried on so successfully that in 1816
the town was second to Dundee in the staple industry.
Before 1740 Montrose was distinguished chiefly for its
shipping, but about that time it entered the linen trade,
which rapidly became of great importance.   Its annual
market was long the principal one in the county for linen
yarn, which was brought from all centres within Forfarshire and even from beyond its limits.

The linen industry was fostered by government boun-
ties till 1832, by which time it was thoroughly established.
While linen manufacture has been on the whole steadily
and even rapidly progressive, at times it has undergone
great depression, as in 1826 and 1847.   But perhaps the
most critical period was 1830–40, when cautiously and
tentatively the manufacturers of Dundee began to spin
and weave jute.   Even after it was finally adopted as the
staple trade, and after machinery suited to the working of
the new fibre had been perfected, the industry was ham-
pered because the jute was imported through London
and Liverpool.   In more recent years a fleet of magnificent
ships has brought the raw material direct from India to
Dundee.   Foreign competition, and especially that of
Calcutta, has operated adversely.   Nevertheless the ever-
increasing markets opened up in nearly every country of

the globe and the fact that certain classes of goods can still be best made in the old centres of manufacture have enabled Dundee to more than hold its own.

Industries auxiliary to spinning and weaving, the main branches of the linen trade, are bleaching, dyeing, and calendering. The bleach-field is a necessary adjunct of every factory or of every textile district. In no part of the county, however, is bleaching so characteristic an employment as in the valley of the Dighty, doubtless because of its proximity to Dundee. This rivulet not only supplied water for bleaching but in the beginning of last century was spoken of as the hardest worked stream in Great Britain in proportion to its size, for there were then on its banks nine bleach-fields, 17 mills for washing and cleaning yarn, and five mills for beating thread and cloth.

The following excerpt from the latest annual report of the Dundee Chamber of Commerce contains interesting figures in connection with the local staple trade :

The local imports of jute for the year ending 31st December, 1911, amounted to 201,000 tons. The imports of flax were 12,600 tons, tow 4250 tons, and hemp 2500 tons. The exports of jute yarn from the United Kingdom for the year 1911 amounted to 22,020 tons, valued at £704,000. Jute piece goods were exported to the extent of 149,450,000 yards, valued at £2,045,000. Jute sacks were exported to the extent of 60,153,000 sacks, valued at £1,051,400 in 1911. Jute manufactures were imported from abroad to the value of £2,163,000, and were re-exported to the value of

£1,324,000, leaving a difference of £839,000 as the apparent value of foreign jute manufactures retained for home consumption.

While in a general sense Dundee is the centre of the linen manufacture, and while in Forfarshire linen is still a great industry, the manufacture of linen is small in comparison with that of jute. Dundee is emphatically "juteopolis."

At one time Dundee had cotton manufactures, and had a wide reputation for knitted bonnets: indeed one of its districts was known as "the Bonnet Hill." Certain streets in the city still retain names associated with other trades—the refining of sugar, which was continued until about 1830, the making of soap, and the manufacture of glass. Tanning, the preparation of leather, and the making of boots and shoes were extensively carried on, though at present these are in abeyance, in Dundee at least. Arbroath, however, still does a large trade in goods of that class. At one time there were no fewer than 60 master-brewers in Dundee, and brewing is still carried on. The export whiskey trade is very extensive In Dundee, Montrose, and other centres, a considerable trade is done in flour-milling. Sweets and preserves are made in enormous quantities in Dundee, its marmalade being known all the world over.

Dundee has for centuries been noted for ship-building; and even in these days when iron ships have all but superseded those made of wood, vessels intended for use in the Arctic and the Antarctic, for whale-fishing and for exploration, are largely made in its yards.

Jute Warehouse at Wharf, Dundee

Of the finer kinds of timber imported, much is used in Dundee for fitting out passenger steamers. But a striking feature of the local timber trade is its extensive case and box manufacture. A large proportion of the wood imported is of course used in the building trade. Paper is manufactured on both sides of the Tay, and esparto grass, wood pulp, and other necessary materials enter the district through Dundee and its sub-ports.

Next to the staple trade the manufacture of iron goods and machinery is the most important in Forfarshire. Practically all the machinery required for preparing, weaving, and finishing jute textures has all along been made in Dundee. The normal production of jute looms in this centre is from four to five thousand annually. The bulk of these is now exported to Calcutta, but large numbers are also sent to continental countries, Japan, China, and Argentina. The value of textile machinery manufactured here approximates to £300,000 per annum, and from three thousand to four thousand hands are employed in the industry.

Dundee engineering firms also make marine engines, boilers, forgings, and castings; besides which there are such specialities as high speed gearing for electric transmission, and machine-cut wheels. This has grown to such an extent that one firm is now recognised as amongst the largest makers of this class of gearing in the world. Another important industry in Dundee is the production of linoleum machinery for local, continental, and American factories. Machine making is also extensively carried on in Arbroath.

Dundee is a very important financial centre. Indeed it is questionable whether the capital of its great investment companies, which do business mainly with America, is not greater than that devoted to its staple trade.

Forfarshire does its own share in printing and publishing. Some of its newspapers have a circulation almost co-extensive with Scotland. Dundee possesses one of the largest photographic publishing businesses in the world.

14.   Minerals—Quarries.

Forfarshire cannot be said to be rich in minerals, though it possesses quarries of exceptionally fine building stone and paving stone. Certain older writers, indeed, speak of mines and of veins of ore that seemed to them to give promise of great things; but these expectations have not been realised.

In the reign of James V, Robert Seton writes, "Some report that at Clovo, at the head of South Esk, some eight miles from Killiemuir, there is found gold and silver." And in his description of the county written in 1678, Edward says: "As to the metals contained in the bowels of this county it is affirmed that different kinds of them are to be found in the valley of North Esk. The great-grandfather of the present proprietor of Edzell discovered a mine of iron at the wood of Dalbog, and built a smelting-house for preparing the metal. This gentleman's grandson found some lead ore near Innermark, which he refined. The son of this

latter found a very rich mine of lead on the banks of
the Mark, and about a mile up the valley from the
castle of Innermark. In a mountain of hard rock where
eighteen miners are digging deeper every day, they have
come to a large vein of ore, which, when the lead is ex-
tracted and properly refined, yields a fifty-fourth part of
silver. This vein seems to be inexhaustible." Elsewhere

A Forfarshire quarry

we read that Sir David Lindsay of Edzell in 1593 discovered in Glen Esk two mines of copper.

But to pass from the experiments and too often
unfulfilled expectations of former days, the latest Government returns of quarrying in Forfarshire will show us the
nature and the extent of modern operations. The figures
are :—

Clay	3,100 tons
Gravel	340 ,,
Igneous Rock	73,326 ,,
Sandstones	52,875 ,,
	129,641 ,,

The making of bricks, a relatively insignificant industry in the county, is carried on at Barry, Montrose, and elsewhere. Whinstone and other igneous rocks are used mainly for road-making, and are quarried all over the county. The most important sandstone quarries are in the neighbourhood of Dundee and Arbroath. Farther north many quarries have in recent years been closed either because they are exhausted or cannot be profitably worked, or because, in paving, concrete has largely superseded stone.

The total output of sandstone, quartzite, etc. in Scotland in 1910 was 743,189 tons; and the following table shows the number of people employed in the chief districts, the production in tons, and the values for the various counties :—

County	Persons Employed	Output	Value
Dumfries	595	102,497 tons	£44,953
Lanark	998	97,766 ,,	30,967
Fife	1063	255,078 ,,	30,027
Forfar	518	52,875 ,,	23,100

From this it would appear that while the output from the Forfarshire quarries is comparatively low, the value per ton is relatively very high. Indeed Forfarshire building stone and paving stone have long enjoyed a high

reputation throughout the country, and many important
edifices public and private in Edinburgh, Glasgow, Dundee,
and other towns have been built of stone obtained from
Carmyllie, Leoch, Pitairlie, Duntrune, Wellbank, West-
hall, and other Forfarshire quarries.

The Forfarshire stones are sandstones, varying in
texture and in colour, as blue, grey, pink, and red.

15. Fisheries.

Fish are caught off all our Scottish coasts, but the
east is far more important than the west, partly because
the North Sea is one of the finest fishing grounds in the
world, and partly owing to the remoteness from markets
of the north-western and Hebridean ports. The relative
importance of the coasts in the fishing industry for 1910
will be seen from the following figures:—

East Coast—total value of all fish landed			£2,225,170
Orkney and Shetland	,,	,,	606,442
West Coast	,,	,,	338,535
		Grand total	£3,170,147

The value of the shell fish caught on the West Coast
exceeded, however, that on the East by £13,617.

Under the Fishery Board of Scotland the whole
country is divided into a number of districts, each under
the supervision of a fishery officer. Each district includes
a number of "creeks" or subsidiary ports. The "creeks"
of Forfarshire and a few in Kincardineshire are in the
Montrose district, those in our county being Dundee,

Broughty Ferry, Westhaven, Easthaven, Arbroath, Auchmithie, Usan, Ferryden, and Montrose. The chief kinds
of fish landed in Forfarshire in 1910 arranged in order of
market value were haddock, cod, herring, sole, plaice,
whiting, mussels, turbot, crabs, which, with less important
varieties, reached a total value of £71,382.

Forfarshire fishermen prosecute their calling in the

Sprat Fishers in Dundee Harbour

Firth of Tay, where salmon, sprats, sparling, etc. are
taken; on the ocean foreshore; but chiefly far out at sea,
where operations are carried on from 16 to 80 miles in a
south-easterly direction, the principal fishing-grounds being
from 40 to 60 miles off shore.

In 1910 Dundee showed an increase of over £5000
in the value of fish landed by trawlers, the chief kinds of

fish brought into the port being haddocks, codlings, whitings, flat-fish, and sprats. Though not the headquarters of the Forfarshire fisheries, Dundee receives half of the entire "take" of the Montrose district.

The value of shell-fish landed at Broughty Ferry far exceeded that of other ports in the district, but in 1900 for the first time in living memory no boats were fitted out for the summer herring fishing. The fishermen of Westhaven and Easthaven are chiefly engaged in the capture of salmon, lobsters, and crabs. Arbroath occupies the third place amongst the fishing stations of Forfarshire, and Montrose the second, although the two ports taken together land only one half as many fish as are brought into Dundee.

While the fishing industry of Scotland increases year by year, some significant changes are taking place in the number of vessels engaged and in the methods of capture adopted. In 1908 the entire fishing fleet consisted of 11,576 craft of all kinds with a value of £2,029,384; but in 1910, when the latest report was issued, the number of boats had diminished to 9724, while their value had increased to £5,439,857. The reason is that smaller vessels propelled by sail or oar are rapidly giving place to larger and more valuable ships with steam or motor power. Nevertheless, 83 per cent. of fishing boats are still of the older type.

Though these changes are fraught with good to the whole community, there are certain aspects of them that are disadvantageous. Year by year it is becoming harder for owners of the small kinds of craft to compete with

larger, more expensive, and more effective vessels owned
by wealthy companies; and, moreover, relatively fewer
men can now find employment. The result is a tendency
for unemployed fishermen to leave the picturesque fishing
village with its brown sails and its yellow sands, and seek
work in the larger fishing ports or in great centres of
population, where they but increase the industrial pressure. Hence there is a movement on foot to obtain state
aid for fishermen that will enable them to acquire motor
boats and so compete with mechanically propelled vessels
of greater bulk.

Occupations ancillary to the fisheries are chiefly those
in connection with the curing and packing of herrings
and certain kinds of white fish. There is employment
for a whole army of curers, coopers, gutters, packers,
clerks, labourers, carters, and hawkers. In 1910, 2525
people were engaged in the fishery district of Montrose.

In addition to the estuarine and inshore capture of
salmon, the North Esk and the South Esk have valuable
stations near their mouths. Both salmon and sea trout are
caught further up these rivers, while their upper reaches
and the numerous other streams and the lochs of the
county afford ample sport to the angler for trout and char.

Dundee has long been distinguished as the headquarters
in the country of the whale and seal fishing which used to
be prosecuted chiefly in the waters of Greenland. In the
earlier days of the industry, whale fishing was carried on
in sailing vessels, and about nine of these left the port
annually for the Arctic regions. An important advance
took place in 1858, when steamers were employed. At

first only five steamships sailed from Dundee, but ten years later their number had increased to eleven. The most prosperous year in the history of the Dundee seal and whale fishing was 1874, when eleven vessels captured 44,087 seals and 190 whales, yielding 1420 tons of oil and 1436 cwts. of whalebone, the whole valued at £106,500.

The Whaler " Balaena "

Seal fishing is now regarded as unprofitable, and whaling (with far less satisfactory results than formerly) has been continued only on account of the high price obtained for whalebone. In 1874 this commodity was worth £750 per ton, but as much as £2900 per ton has been paid for it, while the price, which fluctuates according to supply and demand, ranges from £1400 to £2500.

Whale oil, which used to fetch £50 per ton, has fallen to £20, owing to the introduction and ever-increasing use of mineral oils.

The Dundee whaling fleet has in consequence of these changes diminished to eight vessels. The gradual extinction of the whale and the difficulty experienced in capturing a sufficient number to pay expenses will no doubt in time cause the trade to die out altogether.

The preparation of oil and the tanning and dressing of sealskins are industries proper to Dundee in connection with these fisheries.

16. Shipping and Trade.

Forfarshire has three ports, Dundee, Montrose, and Arbroath. Dundee ranks third amongst the ports of Scotland.

Previous to the discovery of America the east coast of Scotland, as facing the Continent, was far more important in its shipping than the west. By 1354 shipping at Dundee had grown to such dimensions as to require special officials to collect the shore dues and customs. In 1447, James II issued letters patent authorising dues to be raised on vessels frequenting the port so that its harbour might be kept in good repair. In those days French and Rhenish wines seem to have been largely imported at Dundee. In 1654 Dundee owned 10 vessels; in 1691, 21; and in 1792, 116, of which 78 were coasters, 34 foreign traders, and four whalers. Between 1700 and 1814 the harbour was managed by the Town

Council, who instead of applying the dues to its improvement, devoted within that period nearly £30,000 of harbour money to general town purposes. In 1815 the First Harbour Act transferred the management to the Harbour Commissioners until 1836, when the Harbour Board was permanently constituted on a popular basis.

Meanwhile improvements and extensions on plans drawn up by Telford were being carried out at a cost of £90,000, and the Graving Dock and the King William IV Dock were opened. In 1832 another—the Earl Grey —dock was added. The 16¼ acres of dock accommodation thus secured sufficed for the coasting trade until 1865. In that year the Camperdown Dock was constructed and the Victoria Dock in 1869, which increased the acreage of the harbour by 19¼ acres. The main reason for the construction of these new works was the direct importation of jute from India; and to-day, such has been the increase of the tonnage of the jute fleet, these spacious basins are quite inadequate. Besides a special Fish Dock for the reception of trawlers and other fishing vessels, 2800 feet of deep river wharves have been built where the largest jute steamers can lie and discharge at any state of the tide. Since these arrangements were completed the measurement of vessels entering the port has increased by more than 100,000 tons, and the Commissioners have now in view a further extension of the harbour. Its present dimensions are :—

Area of docks	38½ acres.
Quayage	17,875 feet.
Shed accommodation	55,889 sq. yds.

Dundee Docks

The following figures show the number and tonnage of
vessels entering the harbour in 1911 :—

Vessels	Number	Tonnage
Foreign	426	378,543
Coastwise	2,047	320,641
River trade	1,507	54,950
Total	3,980	754,134

The steamers carrying jute are about 50 in number and
their average tonnage about 6000.   There is regular

Torpedo Craft at Dundee

communication with Newcastle, Hull, London, and
continental ports.

The harbour is equipped with hydraulic discharging
appliances. The system of using light portable jiggers
can be seen in no other port in the world. The coal

hoist, 70 feet high, is the largest on the east coast and can ship 200 tons per hour.

The strategical importance of the Tay has recently been recognised by the Admiralty, who have formed a submarine dépot at Dundee. Nearly a score of naval vessels, submarines, and torpedo boats have made the port their headquarters.

The revenue and the expenditure of the harbour of Dundee have steadily increased: in 1858 they were respectively £25,045 and £21,544; in 1909, £80,420 and £73,733. The following tables of imports and exports give the best idea of the trade of the port and city :—

Imports into Dundee in 1907.

Corn	£13,844	
Jute	3,990,106	
Metals	24,107	
Wood goods	385,644	
Other articles	1,293,581	
		£5,707,282

Exports from Dundee in 1907.

Jute goods	£799,804	
Machinery and vessels	81,651	
Spirits	6,765	
Whalebone	5,728	
Other articles	76,321	
		970,269
Total Value of Imports and Exports		£6,677,551

Montrose is a very old port. In 1330 Sir James Douglas embarked there, carrying with him the heart

of the Bruce. The annual tonnage of ships entering
Montrose is about 80,000 and that of ships leaving the
port about 30,000.

The imports and exports for 1906 are as follows:—

Imports.

Flax	£266,901
Wood	34,457
Oilseed cake	7,813
Straw	4,661
Wheat	700
Other articles	10,360
	£324,892

Exports.

Fish, herrings, &c.	£11,679
Ships and boats, new, with machinery	8,000
Potatoes	2,748
Grain	1,432
	23,859
Total Imports and Exports	£348,751

In connection with these figures some of the fluctuations
in trade for the four last years should be pointed out. In
1910 the importation of wheat rose to £5590, while that
of oilseed cake fell to £2321; amongst the exports, ships
and boats rose in 1908 to £27,052, while no return was
made for them in 1910. The number of barrels of herring exported from 1906 to 1910 were for the successive
years 9141, 15,364, 8988, 2142, 11,758.

The original harbour of Arbroath was superseded in
the eighteenth century by one more commodious a little

to the west, which in turn was enlarged and improved about 1844 at a cost of £50,000. Since then the entrance has been deepened and a wet dock constructed.

The imports and exports for 1907 are as follows:—

Imports.

Flax	£148,944	
Hemp	7,598	
Wood	6,113	
Oilseed cake	3,350	
Other articles	816	
		£166,821

Exports.

Corn, grain, etc.	2,788	
Fish, herrings, etc.	3,178	
Other articles	666	
		6,632
Total Imports and Exports for 1907		£173,453

17. History of the County.

Numerous Roman remains, more conspicuous indeed a century ago than they are now, prove that the Romans visited various parts of Angus. Camps at Meikleour and Lintrose in Perthshire seem to have been directly connected with others at Coupar Angus, Cardean, Battle Dykes near Forfar, and War Dykes north of Brechin; while former traces of a camp at Cater Milley (*quatuor milia*), beside Invergowrie, and an entire entrenchment at Haerfaulds near Kirkbuddo, point to the possible existence of another Roman military route between the Sidlaws and the Tay.

Within a couple of centuries after the Romans left, Christianity was introduced into Scotland. Tradition has it that as early as 431 the first church north of the Tay was founded at Invergowrie. Boniface, a missionary from Rome, seems to have been there in the seventh century, while in the eighth, St Drostan, abandoning his bishopric in Donegal, settled as the apostle of the faith in the wilds of Glenesk. At a still later period the Culdees established themselves at Brechin.

Forfarshire was one of the main theatres of the long-continued conflict of the Picts with the Scots on the one hand and with the Angles of Northumbria on the other. Historians are agreed in identifying the site of the battle of Nechtan's Mere with Dunnichen in Forfarshire. For long the Angles had assumed possession of Pictland, but on that battlefield their king, Egfrith, was decisively repulsed, and the limits of Northumbria were pushed south to the Tweed.

But the Picts were destined to find a more formidable enemy in the Scots. On the conflict with them as well as on the civil wars amongst the Picts themselves, history sheds an uncertain light. That the Scots were the ultimate victors in their struggle with the Picts is certified by the fact that in 844 both accepted Kenneth MacAlpin as their king; but it is clear from the records of such engagements as the battle of Liff, that their victory was by no means assured from the first. On that occasion, Alpin, a Scottish king, whose headquarters had been at

Dundee, was defeated and slain by the Picts at Pitalpin (i.e. grave of Alpin) near Camperdown.

Angus suffered with other districts from the raids of the Vikings. In 980 they are said to have taken the town and castle of Montrose. In 1012 they appear to have landed in three bands at Montrose, Lunan Bay, and near Carnoustie. They succeeded in burning Brechin, but met with signal defeat at Barry and Aberlemno. Near the former place, in what is sometimes known as the Battle of Panbride, Malcolm II won a great victory over Camus, general of King Sweno of Norway. This leader is said to have been buried at Camuston, where an abnormally large skeleton with part of the skull cut away was reported to have been found. A rude clay urn and a bracelet of gold preserved at Brechin Castle are regarded as relics of this incursion. It is at least certain that in no part of Angus have so many traces of ancient sepulture been found as near Aberlemno and Carnoustie. Numerous traces of the Vikings exist near Lunan and Inverkeillor.

The eleventh, twelfth, and early thirteenth centuries, though comparatively obscure, were a momentous era for all Scotland and are especially noteworthy in the history of Forfarshire. The county was then often favoured by the presence of royalty, and the county-town attained an eminence it was not destined to hold permanently. The names of Malcolm II, Malcolm Canmore and Queen Margaret, David I, William the Lion, and Alexander II, are often closely associated with Angus in such matters as the building of fortalices, the holding of royal councils, the founding of ecclesiastical institutions, and the granting of charters and trading privileges to towns. These also are the centuries during which many aristocratic families settled in this part of the country.

Angus figures prominently in the War of Independence. In the years 1295–7 Dundee changed hands no fewer than three times, Wallace and Scrimgeour, who acted for him in his absence, taking it from the English, and Morton and Edward himself recapturing it from the Scots. At Stracathro in 1296 John Baliol appeared, stripped of his kingly ornaments, before Edward, and formally surrendered all claims to the kingdom of Scotland. In 1306 the National Council met in Dundee and declared Bruce rightful king. Six years later his brother took Dundee from the English. Five times between 1320 and 1328 Bruce took up his quarters at the Abbot's house in Arbroath; while parliament met in the abbey and issued the Declaration of the Independence of Scotland, and their remonstrance against the excommunication of King Robert by Pope John XXII. During the minority of David II, his regent, Sir Andrew Murray, gained a notable victory at Panmure over Lord Montfort commanding an army for Edward III.

The Grampians formed the natural boundary line between Celtic and anglicised Scotland, and Angus, lying as it did just within the borders of the latter, had to bear the brunt of many an invasion from the north. About 1392 the son of the Wolf of Badenoch made his appearance amongst the Braes of Angus at the head of a band of Highland caterans, and wasted the country. Walter

Ogilvy, the sheriff of the county, defeated him with great slaughter.

In those days family feuds were common. Ogilvy of Inverquharity had superseded the Master of Crawford as justiciar in the regality of the Abbey of Arbroath. This caused a fierce feud between the Lindsays and the Ogilvys; and in the battle of Arbroath, in which both sides suffered severely, the latter were defeated. Inverquharity is said to have been smothered in the Castle of Finhaven.

A mysterious affair in the reign of James V was the execution of Lady Glamis, sister of the Earl of Angus. Having "conspired and imagined the destruction of the most noble person of our most serene lord the king by poison," so ran the charge against her, she was burned to death upon the Castle Hill, Edinburgh, "with great commiseration of the people in regard of her noble blood, being in the prime of her life, of a singular beauty, and suffering, although a woman, with a manlike courage."

In the Reformation Forfarshire played an important part. Indeed, Dundee was the first Scottish burgh to declare for the reformed faith, and the inhabitants signalised the new departure in 1543 by destroying the houses of the Black and the Grey Friars.

With the rest of the county, Forfarshire suffered alike from English and French attempts at domination during Queen Mary's minority. Dundee was taken and fired by the English soon after the battle of Pinkie. Broughty Castle, which the English seized about the same time, was recaptured by the Scots and the French in 1549.

The next decade is marked by the ascendency of the
Lords of the Congregation and their struggle with the

James Graham, Marquis of Montrose

queen-regent, Mary of Guise. Dundee was for some time
their headquarters. They seized and fortified Broughty
Castle.

When James VI left the kingdom to succeed Queen Elizabeth, deadly feuds raged amongst the nobles. In Forfarshire Ogilvys and Lindsays carried on their ancient family animosities. These feuds were not always confined to neighbouring families. The burning of the "Bonnie Hoose o' Airlie" was an act of vengeance on the Ogilvys carried out in 1640 by the far-stretched arm of Argyle.

During the campaign of 1644–1645, the Marquis of Montrose did not spare his native shire. Covenanters and Royalists fought in the town of Montrose; and next year the Marquis stormed and pillaged Dundee, which supported the Covenanters.

Between 1645 and 1648 Forfarshire suffered along with the rest of the country during a visitation of the plague which is said to have carried off nearly half of the inhabitants. The attempt made by Charles II, in 1650, to escape from the Covenanters, which is known in history as the "Start," had its scene mainly in Angus. Taking flight from Perth, he rode by way of Dudhope, Auchterhouse, and Cortachy to Clova, where he was overtaken by his pursuers "in a nasty room above a mat of sedges and rushes, overwearied and very fearful."

Next year General Monk captured and sacked Dundee. In 1689 Forfarshire was the scene of several of Claverhouse's movements and operations; and in 1715 it was strongly Jacobite. The magistrates of Dundee being for James, the Old Pretender, although most of the citizens were Hanoverian, Graham of Duntrune proclaimed him king there. On January 16, 1716, James,

John Graham of Claverhouse

accompanied by his lieutenant, the Earl of Mar, entered
Dundee; and, on the collapse of the rebellion, he sailed
from Montrose. In 1745 there was strife at Montrose
between the Jacobites and the Royalists. Dundee was
seized and held for five months for Prince Charles. After
Culloden many of the Jacobites found refuge in Glenesk;
and Dundee being recovered for the Royalists, Cumber-
land entered the city and received in a golden casket
a free burgess ticket.

Towards the close of the eighteenth century the bread
riots of "the Meal Mob" occurred in Dundee; and a
celebrated local incident was the planting in the High
Street of the "Tree of Liberty," a piece of work so
distasteful to the provost that he had the tree uprooted
and thrown into a cellar, whence it was afterwards
removed to a garden in the west end of the city. As
might be expected the French Revolution was not
without its effects in a district so radical. Political
discussion was rife and the authorities put it down with
a high hand. George Mealmaker's address to the people
of Scotland on the subject of reform was edited by a
Dundee clergyman, Thomas Fysche Palmer, who paid for
his fearless efforts in the cause of liberty by seven years'
transportation to Australia, in which he was accompanied
voluntarily by James Ellis, an ardent admirer of his action.
In 1816 Parliamentary reform was actively taken up by
Dundonians, and is associated with the names of Rintoul,
Mudie, and Kinloch. Local interest in the subject is
memorialised in street names, like Reform Street.

18. Antiquities.

In cave-dwellings and in tombs there have been found throughout Scotland arrow-heads, spear-heads, and axes made of flint, the weapons of the men of the Stone Age. The Bronze Age followed but did not at once supersede the Stone Age. When the Iron Age came, the Caledonians would have abundance of material within their own land for their uses, though the puzzle is how they could produce the intense heat necessary for smelting iron ore. As in the case of the earlier periods, it must not be assumed that bronze or even stone weapons at once became obsolete on the discovery of iron. Various places in Angus have yielded swords, daggers, spear-heads, besides implements and personal ornaments.

In the conflicts between the Romans and the Caledonians, the invaders seem to have employed two routes through the county. The main route ran along the middle of Strathmore ; the other, which would secure communication with the sea, began on the Firth of Tay and skirting the eastern heights of the Sidlaws joined the central road some distance from the point at which it entered the Mearns. There are remains of the former road, now nearly obliterated, in the parish of Airlie, and there was a camp at the junction of the Dean and the Isla. Another situated about a half-mile north-east of Forfar, was capable, according to the estimate of experts, of containing 26,000 men. In the parish of Oathlaw, at Battle Dykes, are the remains of another camp, stated to be three times as large as the famous camp at Ardoch. Yet

another camp on the main line of communication through
the county had its site at War Dykes, or Black Dykes, three
miles north of Brechin. The chief camp on the southern
route was probably that of Haerfaulds, near Kirkbuddo.
It measures 2280 by 1080 feet and is believed to have
been capable of holding 10,000 men. The two routes
may have had their junction at Aesica, which was
probably situated on the South Esk. Time, the utilitarian
ignorance of early builders, and agricultural improvements,
have removed most of these Roman remains; and in
particular all traces of the southern route have been
obliterated. Even as late as the end of the eighteenth
century many of these relics were much more easily
seen than they are now. Fortunately measurements of
camps and ramparts that have now disappeared were made
in time to record their existence. An aureus, or gold
coin, of the reign of Antoninus Pius was found in the
parish of Kinnell in 1829.

Of the mysterious monoliths, often called Druidical
stones, Angus contains many examples. When the stone
monument consists of two or more uprights supporting a
horizontal slab, it is styled a cromlech. This is usually
found to be a place of burial. Cromlechs are rarer than
the single stones; and it is therefore specially noteworthy
that there is a fine specimen on the Sidlaws in the parish
of Auchterhouse. Another curiosity of the same class is
the rocking-stone, a huge block delicately poised on
another, so that it may easily be set in motion without
being dislodged. There used to be specimens at Gilfum-
man in Glenesk and on Hillhead, Kirriemuir.

Besides such gigantic memorials there are numerous other sepulchral remains—the cairn, the barrow, the tumulus, the mound—all of them monuments of the honoured dead. These, it appears, are found neither in mountainous districts nor in carse lands, but in pastoral regions of the centre and the north. In Forfarshire such primeval cairns are found in most parishes. At one time it was enough to open one of these to find relics of the ancient dead.

Weems are specimens of very primitive architecture. They are built against the slope of some hill or under flat dry ground, but so completely covered up as to make their discovery accidental. A low entrance which must be crawled through leads by a crooked passage to a chamber dark and airless but for a small aperture or chimney in the further end. The floor slopes down until it is possible in some cases for a man to stand upright, though in others the height of the structure is only five feet. One of the most perfect of these weems, or Peghts' houses—perhaps the best in the kingdom—was found on the farm of Barns in the parish of Airlie. It is 70 feet in length. Another exists in the face of a brae at Ruthven. In a weem accidentally discovered in 1871 in a field at Tealing were found horses' teeth, a piece of Samian ware, a bracelet, bronze rings, pieces of burial urns, and ten querns, or hand-mills for grinding corn. These by no means exhaust the instances in Forfarshire of these dismal but interesting abodes.

The county, too, has its specimen of a lake-dwelling or crannog. A crannog was an island, wholly or partially

Weem or Earth House at Pitcur

artificial, situated in a loch and sometimes joined to the
shore by a causeway that might easily be submerged or

Jet necklace found in cist at Tealing

destroyed. Such structures, to judge by remains found
in them, were not merely fortresses but dwelling-places
for one or more families. The first Scottish crannog to

be brought into prominent notice was discovered in the Loch of Forfar when it was partially drained in 1780.

The kitchen-midden already mentioned as having been discovered at Stannergate near Dundee contained no fewer than twelve stone cists, eight the full length of a human body, and four shorter ones in which the corpse had been doubled up. In one cist was a coarse urn. When the Dundee and Arbroath railway was being made more than half a century ago, other cists were found in the same neighbourhood, showing that the place must have been an ancient grave-yard. Twelve feet under the cists, the workmen came upon large beds of shells. Amongst these were two antlers of red deer and a piece of flint. This deposit is probably the oldest of the kind in the district.

Besides such rude sepulchral remains, traces of ancient forts give evidence of the doings of the old inhabitants of Angus. The tops of high hills were used as places of refuge or of observation, and as suitable points on which to kindle warning beacon fires. Hill forts consist for the most part of mounds of earth or stone or both, running round the crests of hills. A remarkable instance was discovered on the Laws, a rocky, wooded hill near Monifieth. The hill itself is composed of trap, but on the top was a collection of freestones, which neighbouring builders long used as a kind of quarry. Having cleared away what remained, the proprietor discovered a vast series of trap boulders so put together as to give evidence of having been foundation walls of some great building. One circular wall about 18 feet thick is pierced by a

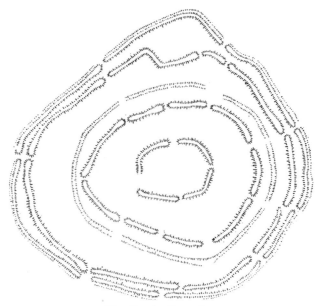

The Brown Caterthun

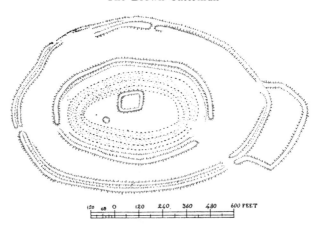

120 60 0 120 240 360 480 600 FEET

The White Caterthun

passage leading into what must have been a round chamber 40 feet in diameter. The débris contained a stone cup, a coin, an armlet, iron axes, an iron sword, and charred remains of bones, wheat and barley. It is now impossible to say whether these elaborate structures were fortress, temple, or tomb. It is significant that they should have been erected on a spot commanding a wide expanse of country.

A famous example of vitrification as applied to forts exists on Finhaven Hill near Forfar. The height of the hill is about 600 feet, and the dimensions of the fort which once crowned its summit are some 400 by 112 feet. There are traces of internal divisional walls. Many parts of the walls are vitrified, and contain various kinds of stone fused together by the external application of heat. This is commonly supposed to be a Pre-Roman fortification.

By far the most notable instances of hill forts, whether in Angus or in the whole country, are those of the Caterthuns in the parish of Menmuir. Two conjectures have been offered as to the meaning of this name, which is of Celtic origin : *cader dun*, hill fort ; and *caither dun*, temple hill. These marvellous structures are on two hills about three-quarters of a mile apart and called respectively the White and the Black Caterthun. The White Caterthun may have got its name from its rings of white stone ; the other, by way of contrast, is black, the dark lines being scarcely distinguishable at a distance from the hue of the heather. The White Caterthun is by far the more important.

The fortress consists of four concentric circles of stone, the innermost of which is about 80 paces in diameter. "The most extraordinary thing that occurs in this British fort is the dimensions of the rampart, composed entirely of large loose stones, being at least 25 feet thick at top, and upwards of 100 at bottom, reckoning quite to the ditch, which seems, indeed, to be greatly filled up by the tumbling down of the stones. The vast labour that it must have cost to amass so incredible a quantity, and carry them to such a height, surpasses all description. A single earthen breastwork surrounds the ditch ; and beyond this, at a distance of about fifty yards on the two sides, but seventy feet on each end, there is another double entrenchment of the same sort running round the slope of the hill. The intermediate space probably served as a camp for the troops, while the interior post, from its smallness, could only contain a part of them. The entrance into this is by a single gate on the east end; but opposite to it there are two leading through the outward entrenchment, between which a work projects, no doubt for containing some men posted there as an additional security to that quarter."

Sculptured stones are very numerous in Forfarshire, but it is hazardous to dogmatise on the date or the origin of these. Some, however, belong clearly to Pagan times, others to Christian.

The pillar-stone at Kirkton of Dunnichen, with its Z-shaped ornament, comb, mirror, etc. is evidently of the Pagan period. Cairn Greg, an eminence on the estate of Linlathen, was found on being excavated to contain a cist, within which were a bronze dagger and an urn with ashes. Above this was a sculptured stone pointing to heathen usages and beliefs. The legend of a dragon-haunted well, where nine maidens are said to have perished, is as old as the stone, near Balluderon, carved with a transfixed serpent and the zigzag symbol. An inscription on a sculptured stone at St Vigeans has been regarded as the sole specimen of Pictish writing that has come down to us. It speaks of the stone as erected to Droston, son of Voret, of the race of Fergus; and a Pictish king, Droston, was killed in battle at a spot a mile or two distant in 729. This stone, which was discovered about fifty years ago, is a broken cross with interlaced tracery, grotesque figures, and a hunting scene in which a man kneeling on one knee is depicted as discharging a cross-bow at a wild boar. On another fragment are priests tonsured in the Roman manner, not in the Scottish or Irish, a peculiarity pointing to the year 710 as its earliest possible date, when Nechtan put his church under St Peter and adopted the Roman customs. An early Christian monument, discovered amidst the foundations of the old parish church of Arbirlot, has on it a cross, two open books, and a small circle. The sculptured stones of Aberlemno have figures of armed warriors.

Stones have also been found at Monifieth, Kettins, Craig, Eassie, Farnell, Kirriemuir and elsewhere in Forfarshire.

19.   Architecture—(*a*) Ecclesiastical.

The monastic institutions of Scotland were founded by David I and his successors in the twelfth and early thirteenth centuries.   Some of the Forfarshire ecclesiastical foundations were peopled from abbeys in the south of Scotland : Arbroath from Kelso, Restenneth from Jedburgh, and Coupar Angus from Melrose.   Arbroath Abbey was founded by William the Lion in 1178.   In 1233 it was dedicated to the memory of Thomas Becket. It was destroyed by fire in 1559.   Long regarded as a quarry, its vast ruins as now preserved attest the extent of the great building and the beauty of its architecture, the First Pointed style, of which it is a fine example.   The western doorway, with a beautiful rose window, now gone, led into the nave, 270 feet long.   In front of the high altar is the supposed grave of the founder.   The gable of the south transept, still fairly entire, has a window, the famous Round O of Arbroath, a landmark seen far out at sea and for centuries a guide to mariners.   The north-west tower, the highest part of the church now standing, rises to a height of 103 feet, but four pillars situated at the point where nave and transepts meet sustained the great central tower, which was probably 150 feet high.   The south wall of the nave is still standing.   An interesting accessory of the building is the Abbot's House, or Abbey House, still entire ; while the Gateway, or Abbey Pend, is a picturesque fragment.

Arbroath Abbey

Of the old religious houses of Forfarshire, the only one still fit to use as a place of worship is Brechin Cathedral. About ten years ago, its ruined parts were rebuilt, its long hidden beauties and graces were restored to view, and the whole edifice was thoroughly renovated. The cathedral was the work of more than one age: its architecture is partly Early English and partly Decorated Gothic and it also exhibits the French flamboyant style. The nave measures 140 feet by 58 feet and is adorned with two rows of pillars and five arches on each side. The western door is elaborately carved, and the graceful mullions and tracery of its windows are very fine. At the north-west corner rises a massive square tower strengthened by buttresses and surrounded by a battlement, the whole surmounted by an octagonal steeple the pinnacle of which attains the height of 128 feet. A smaller tower on the north-east corner of the large tower contains the spiral stair by which the bartizan is reached.

From time immemorial the site occupied by the cathedral has been "a holy place": pagans, Culdees, Romanists, episcopalians, presbyterians have in succeeding ages worshipped there in older fanes or in that which still occupies the ground. Christianity was first introduced there by Columba's missionaries. King Nechtan, who favoured Romish Christianity, expelled the Columban monks in the eighth century; Kenneth III "gave Brechin to the Lord"; and Malcolm II is said to have founded the monastery. David I appointed the first Bishop of Brechin and, as he did not suppress the Culdees in

Brechin, for long there seems to have been the somewhat incongruous combination of a Romish bishop and
a Culdee chapter of monks. No fewer than twenty-three
churches and chapels were attached to this influential
bishopric.

The most interesting ecclesiastical relic in Brechin
is, however, the Round Tower. Long detached from
the cathedral, and how much older than it no one knows,
the Round Tower occupies a site close by on the south-
west; and indeed the two buildings are now united.
The purpose originally served by the Round Tower, its
builders, and its age are puzzles that may probably never
be satisfactorily solved. The only building like it in
Scotland is the Round Tower of Abernethy, but there
are many examples in Ireland. The Brechin Round Tower
is a much more perfect building than that of Abernethy.
Its circumference at the bottom is 47 feet 11 inches, and
at the top 40 feet 10 inches; and the tower proper is
crowned with a tower 18 feet 9 inches high, the total
height of the building being 110 feet. There is an
interesting doorway near the·bottom and facing the west.
The interior consists of seven storeys reached by a series
of ladders.

A mile to the east of Forfar are the remains of
Restenneth Priory, situated on what was an island till
the little loch was drained. The Priory belonged to the
thirteenth century and was a cell of Jedburgh Abbey.
An earlier church is said to have existed here, one of
three—Invergowrie, Tealing, and Restenneth—founded
by St Boniface in the seventh century. Dr Stuart, the

Brechin Cathedral and Round Tower

eminent antiquary, believed the square tower, which at a later period was crowned with an octagonal spire, to be part of the original church.   If this is correct, the church must be one of the oldest ecclesiastical buildings in Scotland.

Dundee possessed several important monasteries, though

Restenneth Priory

scarcely a trace now remains of them.   The Howff, or old burying-ground, was the site of the Grey Friars, the earliest religious house in Dundee.   Opposite the foot of South Tay Street, where the chief Roman Catholic church in Dundee now stands, was the monastery of the Red Friars.   On the west side of Barrack Street (formerly Friars' Vennel), facing what would be the monastery of

the Grey Friars, stood that of the Dominicans or Black
Friars.   Of the Dundee nunneries one was situated
between Bank Street and Overgate, its buildings being
taken down in 1869; and the other, perhaps at the foot
of Step Row, remains in name, at least, in the Magdalen
Green.   Montrose also had a monastery and perhaps a
nunnery, while the existence of several chapels is inferred
simply from the names of lanes and streets.

Many of the modern churches of the larger towns
in Forfarshire furnish examples of various styles of architecture.   The graceful spire (1832–34) of the parish
church of Montrose, 220 feet high, is a striking and
pleasing object in the landscape.   In Brechin two
handsome churches are the Gardner Memorial Parish
Church and the new Maison Dieu United Free Church,
while the early Gothic style is finely exemplified in the
Bank Street United Free Church.   The parish church
of Arbroath, built in 1590 with materials taken from
the Abbey Dormitory, had a handsome Gothic spire,
one of the finest of its kind in Scotland, added to it in
1831.   When the church was destroyed by fire in 1892,
the spire fortunately escaped destruction.   St Mary's
Episcopal Church, a good Gothic building with a spire,
is noteworthy.   The parish church of Forfar is conspicuous by its handsome spire, 150 feet high.   The
English Episcopal Church of St John the Evangelist, in
the Early English style, is one of the finest edifices in
the county town.   Only 40 feet of the spire (its projected height is 163 feet) have yet been completed.   The
parish church of Kirriemuir, a handsome edifice built in

Old Steeple, Dundee

1786, has a neat spire. In Dundee and its residential
suburb, Broughty Ferry, are many fine modern churches.
The "Old Steeple" forms a belfry to a group of three
Established churches, the most easterly of which, St
Mary's, is the parish church of Dundee. Towering
high above all the ecclesiastical edifices of the city and
district, the magnificent Gothic spire of the Cathedral
Church of St Paul's, the seat of the Bishop of Brechin,
rises to a height of 210 feet. It is built on the site of
the ancient Castle of Dundee, and was erected in 1855,
the architect being Sir Gilbert Scott.

20. Architecture—(*b*) Castellated.

Amongst military structures in Scotland, there are
said to be no examples of the Norman style. Our oldest
castles belong to the succeeding period. The typical
fortress of the early fifteenth century was the simple
square tower with three or four vaulted storeys. Such is
Melgund castle. In order to provide for some flanking
work, the builder of these square towers often perched
turrets on the corners or top of the building. There
might be four of these works, but often the poverty of
the laird restricted him to one or two. After the long
quarrel with England, Scottish architects adopted French
rather than English models. Some of the distinctive
features of this Scottish baronial style, for which we are
in part indebted to our French allies, are turrets projecting from the wall upon bold corbellings and terminating

in pointed roofs; towers of circular plan; parapets and battlements; roofs of steep pitch; gables of stepped outline; small, square windows; plain, unadorned doorways; prominent or lofty chimneys.

Fortresses are situated at points of vantage amongst or near the hills, where they have been erected to guard some pass or ford, or at places on the coast liable to attack. Forfarshire, for example, has its castles placed at, or near, the entrances to glens leading far into the Grampians, at passes in the Sidlaws, at important strategic points in Strathmore and Strathmartine, and on or near the shore of firth and ocean.

Situated in Glen Isla at a point to the north-east of Mount Blair, and commanding the route through that glen and one leading over to Glen Shee, Forter Castle, an ancient stronghold of the Ogilvys, is a good example of a mountain keep. It has long been in ruins, but a special interest attaches to it as the actual scene of the burning of what in the ballad is called the " bonnie house o' Airlie." The Earl of Airlie was obnoxious to the Covenanters and devoted to the Stewart cause. During his absence in England in 1640, Argyll as leader of the Covenanters and in prosecution of an old family feud between the Campbells and the Ogilvys, attacked and destroyed both Forter and Airlie. The latter castle occupies a romantic position at the junction of the Melgum and the Isla. The precipices rising from both stream and tributary rendered it impregnable on all but its landward side, which was defended by a deep fosse, drawbridge, and portcullis.

One of the finest ruins of the county, Inverquharity Castle, is situated on the South Esk near the point where that river is joined by the Carity Burn.   It formed another of those defences against raiders down Glen Prosen

Inverquharity Castle

or Glen Clova.   Built by one of the Ogilvys, it is a large, square, battlemented tower, with walls 9 feet thick. The castle possesses one of the few iron gates, or "yetts," met with in the county, for the use of which a special licence from the sovereign was required.   In the

neighbourhood are the ruins of Clova Castle.   Cortachy
Castle at the junction of the Prosen and the South Esk,
has a name that is said to indicate the nature of its site—
"the enclosed ground."   Shut in from the surrounding
world by hills and woods, it is built in an ideally seques-
tered spot.   Like so many others in Forfarshire the castle
consists of portions built at separate periods; but not
even its oldest part, a circular tower terminating with
a square corbelled superstructure, looks as if it had been
a fortress, though in the early sixteenth century, when
it seems to have been built, it must have served to some
extent as a place of defence as well as residence.

V ayne Castle, amidst the Braes of Angus between the
two Esks, is situated on a precipitous rock overlooking the
Noran.   The castle, built of red sandstone, consisted of
three storeys and had a circular tower containing a stair-
case. Probably a stronghold of the Lindsays, it is another
good instance of a fortress built to guard against northern
forays.

Finhaven Castle, whose ruins may still be seen near
where the Lemno joins the South Esk, was for a long
time the chief seat of the Lindsays, Earls of Crawford.
James II defied the notorious Earl Beardie and vowed
he would make the highest stone of the castle its lowest.
He subsequently pardoned the Earl, but to fulfil his
vow—

> " Bounding nimbly to the highest tower,
>   Where Beardie wont to pass his leisure hour,
>   Down to the lawn a crazy stone he threw,
>   And smiling cried—' Behold my promise true!'"

Glenesk, even more closely associated with the "lichtsome Lindsays," contains two interesting memorials of that warlike family. Amid the rocky fastnesses near the head of the glen, rises the ruinous tower of Invermark Castle, the most typical specimen probably

Invermark Castle

in the whole county of a mountain fortress. Built of native granite, the tower is one of four storeys, with the entrance on the second floor, now somewhat difficult of access. In days of yore it was reached by a drawbridge stretching between the castle and a flight of steps 12 feet distant. Should an enemy have succeeded in crossing this he would have found the doorway guarded by a strong iron "yett," within which was an oaken door.

Near the entrance to Glenesk stood Edzell Castle, whose ruins form the most impressive relic in the county of a medieval fortress and baronial residence. It vied with Finhaven as a seat of the Lindsays. The ruins of the castle testify to the greatness of this family. The

Edzell Castle

lofty "Stirling Tower," a keep 60 feet in height, still fairly entire, is the most imposing part of the ruins. The great baronial hall measured 36 by 24 feet. The courtyard was 100 feet in length by 70 feet in breadth. The castle and its gardens covered about two acres of ground. The gardens were surrounded by walls orna-mented with sculptures and decorations, which may still be seen. Its kitchen, "the Kitchen of Angus," was so large that a whole ox could be roasted in it; and need was, for crowds of noble guests and their retainers were constantly received within the hospitable walls of the castle.

Hatton Castle, near Newtyle, was a fortress of the Oliphants, built in 1575 and noteworthy for the size of its rooms and its window apertures. Some little distance over the pass leading from Newtyle to the valley of the Dighty is the mansion of Auchterhouse. In its grounds is the so-called Wallace Tower, with walls 12 feet thick. This has the reputation of belonging to the early twelfth century; and, if so, it is one of the oldest ruins in the county. The association of its name with the national hero is consistent with the tradition that it was for some time occupied by Edward I.

For centuries the chief seat of the Earls of Strathmore has been Glamis Castle, one of the noblest architectural ornaments of Angus, and the finest specimen in existence of the Scottish baronial style. We are here concerned mainly with the older parts of the building, the construction and character of which take us back to the eleventh century. The room is still shown in which, according to tradition, Malcolm II died, in 1033. Another account speaks of a violent death, and certain obelisks with rude carvings in the vicinity seem to bear testimony to this. The truth can never be known; but the king's death being associated with Glamis bears undoubted witness to the antiquity of the castle. The demesne came into the possession of the Lyons in 1372, the founder of the

Strathmore family being Sir John Lyon, who married the Princess Jane, second daughter of Robert II. Weird stories, ghostly and other, cling to the haunted rooms of Glamis Castle. Residence as it now is and long has been, the enormous walls, the small windows, the arched and groined roofs bespeak the fortress, and justify us in ranking Glamis amongst military buildings.

Forfar Castle has now disappeared, but its ruins were in evidence about the beginning of last century. It was an important fortress in the early days when it was captured by Edward I and so obstinately held for him, that on its being reduced Robert I caused it to be destroyed lest it should again harbour his enemies. Its site is now occupied by the tower of the Market Cross. The district that intervenes between Forfar and Brechin contains the castles of Careston, Melgund, Flemington, and Aldbar, in addition to Finhaven already mentioned. Melgund, which resembles Edzell, is perhaps the most interesting of these, and is associated with the name of Cardinal Beaton.

The nucleus of Brechin Castle is as old as the time of Wallace. Edward I occupied it for some time in 1296, and in the following year Wallace took it. Edward now directed a strong force against the Castle, which, after a hot attack lasting for twenty days, surrendered on the death of its brave governor, Sir Thomas Maule. The fortress occupies a strong natural position, around which on one side sweep the waters of the South Esk. In the fifteenth century it came into the possession of the Maules of Panmure, and is now the chief seat of the Earl of Dalhousie.

Almost directly south of Brechin and about halfway between that city and Arbroath, rises to a height of 60 feet the massive square tower of Guthrie Castle. Its age is uncertain, but it seems to date from the fifteenth century, when Sir David Guthrie obtained permission to erect such a fortress-residence as the times required. Its iron " yett " is a noteworthy feature. The old tower is incorporated in the modern castle.

Historically more interesting was the ancient Castle of Kinnaird, whose site is now occupied by what is perhaps the most sumptuous mansion in Angus. The old castle was burned by Earl Beardie in 1452, because its owner fought against him at the battle of Brechin. James VI, Charles I, and Charles II were guests in the building which replaced it ; and Charles I created its owner Earl of Southesk. It was here that the Great Marquis, who had married a daughter of the house, parted from his wife when on his way to execution at Edinburgh. James, the fifth earl, an active Jacobite in the Fifteen, suffered the forfeiture of his title and his estate for his adherence to the Stuarts, and died in exile. In 1855 his great-grandson was reinstated by the House of Lords.

Near Montrose are the castles of Rossie and Dunninald, while to the north-west of Montrose Basin are the remains of the ancient House of Dun, whose proprietors for many generations figured notably in Scottish history. Their rights of harbourage at Montrose brought them into frequent and often serious conflict with the inhabitants of that ancient borough.

Two miles west of Dundee is Invergowrie House, the oldest part of which is believed to go back to the fourteenth century. Of the Castle of Dundee nothing remains but the name. Dudhope Park in Dundee, purchased in 1873 as a recreation ground for the city, contains the Castle of Dudhope, a place of some historical note. The oldest part of the building was erected in 1296 by Sir Alexander Scrymgeour, and being subsequently extended became the seat of the Scrymgeours, Earls of Dundee. The property is described in 1682 as being "ane extraordinaire pleasant and sweet place." It was the abode of the celebrated Claverhouse. In its vicinity he unfurled King James's banner and from it set out for Killiecrankie.

Broughty Castle was founded in 1496, and played many notable parts in history down to 1603. Having long fallen into disuse as a fortress, it was repaired and fortified by the Crown to guard the entrance to the Tay. For this it is admirably suited, being situated on a rocky promontory, once an islet, at the point where the estuary of the Tay narrows to a breadth of one mile. Its guns thus command the entrance of the firth and the waterway to Dundee.

In the district to the north-east of Dundee and Broughty Ferry are several castles, once half fortresses, half mansions. Amongst these the most interesting are Mains, Claypotts, and Affleck, all now in a more or less ruinous condition. Mains, or Fintry, Castle was built about 1562 by Graham of Fintry; and, though long uninhabited, is in a state of wonderful preservation. It is situated on a tributary of the Dighty, and its lofty tower is a conspicuous object in the landscape. It is a good example of the castellated architecture of the sixteenth century. The tower has a penthouse corona. The corbelled abutment of a turret, the arched entrance, the quaint gables are noteworthy features. That it was more residence than stronghold, its situation and its ornate character alike show. Immediately behind Broughty Ferry is Claypotts Castle, of Scottish baronial architecture. It is built on the Z plan, and belongs to the latter part of the sixteenth century. Its oblong keep measures 35 feet by 25. The thickness of its walls and its circular towers bespeak the fortress. Built by the Strachans of Claypotts, the castle came into the possession of the Grahams of Claverhouse, and Viscount Dundee resided for some time within it. The ruin is now the property of the Earl of Home. Affleck, or Auchinleck, Castle, a mile to the west of Monikie, is regarded as a fine specimen of its class. From this castle hints have been taken for the restoration of other buildings of the same type. Like Invermark and a few other Angus castles, this finely built structure had until recently a "yett" or heavy door of grated iron. The lofty square tower, a landmark to mariners, is four storeys in height and has the appearance of a Border peel. The walls are of great thickness and solidity.

About the middle of the curve of Lunan Bay, on an elevated piece of ground, stands the roofless ruin of Red Castle. Erected by William the Lion, it is thought to occupy the site of a still older fortress built to guard the entrance to Strathmore.

21.   Architecture—(*c*) Domestic.

The year of the Union of the Crowns may be taken as marking the change from dwellings designed for protection to those aiming at convenience, health and comfort; and within the seventeenth century accordingly, proprietors of wealth and enlightened taste began to add

Cortachy Castle

to their castles or to transform gloomy fortalices into residences better suited to their requirements.   As the eighteenth century advanced, the love of the classic, or pseudo-classic, in art was modified or superseded by a passion for the romantic and a veneration for the past; and hence proprietors and architects sought to give prominence, in at least the outward appearance of edifices, to those characteristics for which the older parts had been distinguished and to harmonize as far as might be modern alterations and additions with architectural ideals of more primitive times. Of this blending of the new with the old there are many excellent examples in Forfarshire.

Cortachy Castle exhibits the mingling of architecture of different dates—it was not completed till after the middle of last century—but all the parts unite in one superb whole. The interior also is full of interest. The vaulted Charter Room holds the family records, and the "King's Room" was occupied by Charles II on his memorable "Start" from his austere covenanting supporters in 1650. In the magnificent grounds is the "Garden of Friendship," where notable visitors have from time to time planted trees.

To the north of Kirriemuir stands the mansion of Kinnordy, purchased by the Lyells from the Ogilvys of Inverquharity. The old part of the house was built over a century ago and was that in which Sir Charles Lyell, the geologist, lived and studied ; but his nephew, Sir Leonard Lyell, M.P., had it entirely reconstructed in 1880, so that it is now one of the most ornate mansions in the county.

The ancestors of Admiral Duncan, who defeated the Dutch fleet off Camperdown in 1797, purchased the estate of Lundie. Finding the castle that belonged to the estate unsuitable as a dwelling, the first Earl of Camperdown built the modern mansion in 1828, naming it after the scene of his father's great victory. Camperdown House, situated in the midst of magnificent and spacious policies, is built of white sandstone in Grecian style, and has a fine portico supported by noble Ionic pillars. An interesting relic in the grounds is the figure-head of the Dutch admiral's ship, a lion rampant.

Alike on architectural and historic grounds, Glamis Castle is undoubtedly the most interesting edifice of its kind in the county. When Sir Walter Scott first visited Glamis in 1793, the castle had around it seven circles of defensive boundaries; but before his second visit "down had gone many a trophy of old magnificence, courtyard, ornamental enclosure, fosse, barbican, and every external muniment of battled wall and flanking tower"—all this to make the place more *parklike*! There he had seen "the very door from which, deluded by the name, one might have imagined Lady Macbeth issuing forth to receive King Duncan." But mistaken as the "improvers" of the castle may then have been, Glamis is far from being entirely shorn of its antique feudal pomp. "It conveys," writes an expert, "no distinct impression of any particular age, but appears to have grown, as it were, through the various periods of Scottish baronial architecture."

The princely demesne in which it stands is worthy of the venerable pile. Approaching it from the south, the visitor enters by an antique gateway adorned with carved lions, rampant opposant. After traversing a broad avenue nearly a mile long, and for the greater part absolutely straight, one advances to the chief doorway at the base of a quarter-circle tower, flanked at right angles by the two main wings of the building. The great tower, 90 feet high, which forms the central portion of the castle, is crowned with a rich cluster of cone-capped turrets, amidst which are abrupt roofs, stacks of chimneys, and railed platforms. Within the massive door is a heavily grated iron gate, four hundred years old.

Within the castle particularly interesting are the old baronial hall, now the drawing-room, with its pargeted ceiling; the dining-room with its valuable portraits, including one of Claverhouse; and the private chapel adorned with fine paintings by the Dutch artist, De Witt, and carved stalls centuries old. The castle contains a very notable collection of paintings, old armour, richly carved old oak furniture, and other priceless curiosities. In the grounds is a remarkable sundial, 18 feet high and supplied with eighty-four dials. The pedestal is supported by four lions, twice life-size and facing the cardinal points. The view from the tower of the castle is one of the finest in the district, commanding as it does the richly wooded and well-watered Strathmore with its noble environment of mountains.

Panmure House and Brechin Castle are seats of the noble family of the Maules, Earls of Dalhousie, whose extensive possessions, 138,000 acres, make them the greatest landowners in Angus. The former mansion is near Carnoustie. The estate came into the possession of the Maule family early in the twelfth century, and though the Earl of Panmure was attainted as a Jacobite in 1715, the lands were subsequently bought back by the family and the attainder was reduced. The nucleus of the present splendid mansion with its great central tower was erected for the second Earl in 1666, on a spot some little distance from the site of the old castle. It was completed by Earl James, who afterwards suffered attaint. The unfortunate Earl has two memorials in the grounds, one the fine iron gates never opened since he passed through them on his way to exile, the other a tall pillar half a mile distant from the house. Panmure House, as we now know it, was reconstructed when Fox Maule, for some time Secretary of State for War, succeeded to the estates.

Brechin Castle, however, is the chief seat of the Earl of Dalhousie. If devoid of the architectural adornments of Panmure and other homes of the Angus nobility, Brechin Castle, as we have already seen, can vie with any in historical interest. The west front of the castle is its main facade; its centre is characterised by a fine pediment; and the corners of the building are adorned with cone-capped towers. The square tower which forms the highest portion is surmounted by the flagstaff of the Redan fort at Sebastopol, which was presented to Fox Maule, War Secretary at the time of its capture. A fine battlemented wall tops the cliff on whose summit the castle is built and faces the river. In the castle are valuable paintings and a still more valuable library. The exquisite grounds and gardens extend westwards along the banks of the South Esk.

Hospitalfield, near Arbroath, apart from its many intrinsic merits, has the double interest of being an adjunct of the ancient abbey of Arbroath, and most likely the original of Scott's "Monkbarns" in *The Antiquary*. The

Brechin Castle

Hospitium—for such was the first building—is now gone ; but portions of the walls of the barns erected at the requisition of Bernard de Linton, fifteenth abbot, the friend and counsellor of Robert the Bruce, are incorporated in the mansion house. Hence the appropriateness of Scott's name "Monkbarns."   Transformed into a residence, the Hospitium at last became the property of James Fraser,

Hospitalfield House

minister of Arbroath, whose descendant Elizabeth Fraser married the late Patrick Allan-Fraser in 1843. The structure remained unaltered until Mr Allan-Fraser's knowledge and taste made of it the romance in stone and lime it now is. It is Scottish baronial in its every charac- teristic and detail. The lofty square tower with its bartizan is flanked by oblong buildings adorned with oriels and alcoves, buttresses and battlements; and the crow-stepped chimneys and corbelled turrets and the rich ivy with which parts of it are adorned contribute to the antique effect of the whole. The proprietor bequeathed his houses and his means for art education ; and after his death in 1890 Hospitalfield became a residence for art

Kinnaird Castle

students. The house contains rich treasures of art in pictures, statuary, and objects of vertu, and is set in the midst of spacious and artistically laid out grounds.

Situated on elevated ground within a park of 1500 acres bounded on the north by the South Esk, Kinnaird Castle, the home of the Carnegies, with its fine skyline of lordly towers, is a striking object in the wide landscape

that it commands.   Brechin to the north-west, Montrose
and its broad Basin in the east, and even the sparkling
waters of the ocean are visible from its central tower,
which rises 115 feet above the lawns.   The modern
castle is the new-modelled and enlarged reproduction of
the older edifice of 1790.   It is built of a delicate pinkish
freestone characteristic of Forfarshire.   While Scottish
baronial in many of its features, the general lightness and
elegance of its architecture are due to the harmonious
intermingling of French, Italian, and classical forms.   The
structure occupies a square with sides over 200 feet long.
The principal elevation, a magnificent façade of grounded
towers, corbelled turrets, lofty roofs, and spacious oriels,
faces the west.   Separating this front from the great deer
park is a splendid balustraded terrace extending the whole
length of the castle.   A double flight of steps near the
centre communicates between the terrace and a stone
balcony that runs within the projections of the north and
south terminal towers.   If Glamis is the most antique
and interesting of the mansions of Forfarshire, Kinnaird
is unquestionably the most ornate specimen in the county
of modern domestic architecture.

22.   Architecture—(d) **Municipal.**

Montrose, which, in spite of its " gable-endies," has
often been regarded as the most inviting in appearance
of the towns of Forfarshire, has some fine public buildings.
Its broad High Street is certainly one of the handsomest

thoroughfares in the county. Projecting into the street is the town-hall, a large building of four storeys with arcade below and balustrade above, and decorated in front with the armorial bearings of the burgh. Other noteworthy buildings are the Museum and the Academy. Two fine public statues are those of Peel and Joseph Hume, M.P.

Dundee High School

Arbroath possesses examples of fine architecture in its Town-hall, Trades-hall, and Guild-hall. In Brechin the new Municipal Buildings, erected 1894–5, which contain the council chambers and court room, are an elegant pile ; and the Mechanics' Literary and Scientific Institute has long been regarded as the finest specimen of

architecture in the city.   The gift to the community of
Lord Panmure, this handsome Tudor edifice is adorned
with a finely proportioned central tower 80 feet high, and
a castellated parapet ornamented with pinnacles.

Forfar, the county town, has some good public
buildings, of which may be specially mentioned the
new County Buildings, the Town-hall, the Reid Hall,
and the Town Cross, an octagonal turret which marks
the site of the ancient castle.

Apart from the Old Steeple, the ancient buildings of
Dundee, mostly town residences of the county nobility,
are fast disappearing.   One interesting relic of the past,
renovated in 1877, remains in the Cowgate Arch, from
which, during the plague in 1544, George Wishart is
said to have preached to the stricken inhabitants.   Over
the Howff, or ancient burying-ground of the city,
Dundonians keep jealous watch lest its crumbling wall
and interesting old-world monuments fall a prey to the
modern improver.   In its immediate vicinity are the
handsome Post Office and the newly-opened Reading
Room, a particularly graceful building.   Hard by in
Albert Square are the High School, with a fine Doric
portico, the ornate and costly Girls' High School, and the
Albert Institute, a Gothic building.   Adjoining this is
the Royal Exchange and the new Technical College.
The west central part of the city possesses the handsome
group of buildings that form University College, an
affiliated college of St Andrews University.   The Town
House (1743), a venerable building with a fine spire 140
feet high, is far from adequate to modern requirements.

The Court House, like some other public buildings in Dundee, suffers from its site and surroundings. Of the other schools the fine edifice of Morgan Academy is noteworthy. The Royal Arch commemorates the visit

University College, Dundee

to Dundee of Queen Victoria in 1844. An interesting relic of the past is the ancient Town Cross, a slender stone shaft surmounted by a unicorn bearing a shield with the Scottish lion between its fore feet.

23. Communications—Past and Present: Roads and Railways.

The earliest lines of communication through a district must on the whole have followed the main courses of rivers and streams. Old drove roads crossing natural routes transversely often ascended to the top of inter-vening ridges ; but the openings in hills by which side streams enter the main valley must have been the general routes. This is plainly proved by the frequent occurrence of strongholds and afterwards towns at gaps leading into the hills ; the former as a check on raiders, the latter in more peaceful times as markets[1].

The Roman route through central Scotland came into the modern district of Forfarshire near Coupar Angus, passed Cardean, Kirriemuir, and Battle Dykes (three miles north of Forfar), and crossed the South Esk near the influx of Noran Water. Hence it extended to War Dykes (three miles north of Brechin), and at Kingsford crossed the North Esk and entered the Mearns. Traces of this road are still visible in parts, and at one place—between Reedie and Kirriemuir—the modern way coincides with it. This route is punctuated at regular intervals with the remains of Roman camps. Between Cater Milly (near Invergowrie) and Haerfaulds (five miles south-east of Forfar) there probably was a subsidiary route affording the Romans communication with their ships.

---

[1] In districts like Angus a road would also run along the sea shore.

Centuries were destined to elapse before any pathways at all comparable to the Roman roads were constructed. Yet there must have been tracks leading to bridges and fords across the rivers. Until the close of the eighteenth century, the main road north and south through Strathmore crossed the bridge at Brechin.

The constant need for communication between such centres as Dundee, Forfar, Glamis, Brechin, Arbroath, and Montrose would early lead to the formation of beaten tracks, which would afterwards become roads. In Dundee the old roads that led to the neighbouring burghs are still the main arteries. If we remember the importance of Arbroath with its abbey, Brechin with its cathedral, and Forfar with its royal residences, we cannot think of these places as being isolated. In this connection it is interesting to note the existence of the " King's Cadger's Road," a track as broad "as the length of the millwand or rod by which a mill-stone was trundled from quarry to mill." This road began at the fishing village of Usan and ran north-west to Forfar. By means of it fresh fish was daily conveyed to the Court. This royal road was perhaps only a bridle path. At all events heavier articles than fish had to be conveyed through the county on pack horses. For instance, slates and pavement slabs quarried in Glen Ogilvy were carried on horseback to Dundee. In time panniers were superseded by rough sledges dragged over the ground, and these in turn by tumbrils, or carts of an extremely primitive kind. But with the improvement of vehicle, improvement of road had to keep pace.

Forfarshire is now abundantly supplied with excellent

roads. If we consult a cycling map in which the various highways are graded we shall find that the best are (1) the Perth to Aberdeen road, which follows the centre of Strathmore by way of Coupar Angus, Forfar, Brechin ; (2) the road from Dundee to Perth ; (3) the coast road from Dundee to Arbroath, Montrose, and the north ; (4) the road from Arbroath to Forfar and Kirriemuir ; and (5) and (6) those which connect Brechin with Arbroath and Montrose. Roads extending in a northerly direction from Dundee, though of first rate surface, are necessarily hilly : they lead severally to Coupar Angus, Meigle, Kirriemuir, and Forfar. There are many subsidiary and farm-service roads ; and each of the highland glens is supplied with at least one good thoroughfare.

Forfarshire has no canals, though a system of them was projected. In a map published by Robert Stevenson, C.E., in 1819, a canal was planned to begin at the Forth and reach Perth by Dunfermline, Kirkcaldy, Strathmiglo, and Newburgh. Then it was to extend to Coupar Angus, Forfar (with a branch to Arbroath), Brechin (with a branch to Montrose), Stonehaven, and Aberdeen. From this plan Dundee was excluded probably because of its easy means of communication by water with Perth and the coast towns of the adjoining counties. In the early forties of last century Glasgow was reached from Dundee *viâ* Leith and the Forth and Clyde Canal in as many days as it now takes hours to go by rail.

The first railway line in Forfarshire, and one of the very earliest in Scotland, was that connecting Dundee and Newtyle, completed in 1832. Its total length was

The old Tay Bridge, after the accident, 1879

The new Tay Viaduct

$10\frac{1}{2}$ miles. At three steep inclines the train had to be raised or lowered by means of wire ropes worked by stationary engines. For some years horses were employed to haul the trains along the more level portions of the track.

The Caledonian and the North British Railway Companies have important parts of their systems in Forfarshire. The main Caledonian line passes through the heart of Strathmore and throws out branches to Blairgowrie, Alyth, Kirriemuir, Brechin, Edzell, Arbroath, and Montrose. An important line runs from Perth to Dundee and from Dundee *via* Arbroath and Montrose to Aberdeen. A direct line connects Dundee with Newtyle, and another Dundee and Forfar.

The North British line, after crossing the Tay, runs north-eastwards by the coast towns and forms with the Caledonian beyond Montrose a joint line to Aberdeen. The present Tay Viaduct, one of the longest in the world, superseded the first Tay Bridge, which was destroyed during a terrific gale in December, 1879. The new bridge, which carries a double line of rails, forms one of the main links in the east coast route between London and the north.

There is an extensive system of tramways in Dundee and district. The line extends from Ninewells to Broughty Ferry and Monifieth, a distance of seven miles, while to the north-west and north it reaches Lochee and Downfield, the latter a thriving suburb.

24. Administration and Divisions.

In early times shires must have been entrusted to warlike nobles able to maintain order and to levy contributions for the sovereign. The representative of the king in Forfarshire has been named in successive ages maormor, thane, earl, and sheriff, an office which until the middle of the eighteenth century was hereditary in some leading family. In 1747 this hereditary jurisdiction was abolished by act of parliament. That it was an office of emolument as well as of honour is evident from the fact that the sum of £152,037 was voted as compensation to those about to be deprived of it. Of this sum £12,137 fell to the king's representatives in Forfarshire.

After the passing of this act, advocates by profession were nominated as sheriffs and held courts in the county towns. As the population increased and the duties of the sheriff multiplied, sheriffs-substitute were appointed. The sheriff-principal is now entirely an appellate judge. There is one sheriff-substitute for Dundee alone, and another for Forfar and Arbroath. Honorary sheriffs-substitute are also appointed, three in Forfar, seven in Dundee, two in Arbroath, and one each in Montrose and Brechin. The sheriffs-substitute hold small debt courts periodically at Forfar, Dundee, Arbroath, Montrose, Brechin, and Kirriemuir. Justice of the Peace courts, or Petty Sessions, are held in Dundee, Forfar, and elsewhere for small debt and other purposes. The burghs possess police courts presided over by local magistrates to deal with minor criminal offences.

In olden times the sheriff had both legal and military duties to perform; but while the former have been largely extended, the latter have diminished. The sheriff still has, however, the superintendence of the police and certain powers and duties with regard to the military in times of civil disturbance. Under the Territorial Reserve Forces Act, 1907, the Lord Lieutenant

County Buildings, Forfar

is *ex officio* president of the Territorial Association. In Forfarshire there is a Lord Lieutenant, a vice-lieutenant, and twenty-seven deputy-lieutenants; and in the County of the City of Dundee, the Lord Provost is also Lord Lieutenant, and there are nineteen deputy-lieutenants.

For administrative purposes Forfarshire has its County Council, besides which every important burgh has its council for local affairs. The most important, the royal burghs, created by charter from the Crown, are Forfar, Dundee, Brechin, Montrose, and Arbroath. Next come parliamentary burghs, towns possessing the right of sending members to parliament. Police burghs consist of towns with over 700 inhabitants formed under the Police Acts. All three designations may be applicable to one and the same town. In older times there existed also burghs of regality, as Kirriemuir; and burghs of barony, as Glamis and Edzell.

The County Council of Forfarshire consists of fifty-four members (four *ex officio*) representing the various electoral divisions. The fifty elected members are distributed thus:—Dundee District, thirteen; Forfar District, fourteen; Brechin District, twelve; Arbroath District, eleven. The County Council levies rates and borrows money for public works; and has the oversight of roads and bridges, public health, and police. A special District Lunacy Board has the superintendence of asylums.

The Poor Law is administered by parish councils, who also levy rates for primary education. Primary education is entrusted to school boards, of which, apart from towns, there is one in each parish. Special committees take charge of secondary and technical education.

Ecclesiastical affairs in the presbyterian churches are managed by various presbyteries within the synod of Angus and Mearns. Forfarshire has fifty-five civil and thirty-one *quoad sacra* parishes.

Forfarshire was rendered more compact by the Boundary Commissioners in 1892. Before that date there were certain detached portions of parishes within the adjoining counties, and some detached portions of other counties in Forfarshire. Thus the parishes of Alyth and Coupar Angus are now wholly in Perthshire; while the parish of Liff, Benvie, and Invergowrie, and that of Fowlis Easter have passed entirely into Forfarshire.

Previous to 1832 Dundee joined with Forfar, Perth, Cupar-Fife, and St Andrews in returning a member to parliament. Dundee has now two members, the Montrose burghs (Montrose, Arbroath, Brechin, Forfar, and Bervie) one, and the rest of the county one; so that in all Forfarshire is represented in parliament by four members.

25. Roll of Honour.

In the state and the church, in the army and the navy, in letters and law and science, and in the struggles for religious and political liberty, there is many a noted name associated with our county.

The name of Scrymseour or Scrimgeour, "skirmisher or hardy fighter," was bestowed by Alexander I on Sir Alexander Carron, for bravery in assisting to put down a rebellion, and in particular for carrying the royal standard across the Spey. The honour of being standard-bearer was made hereditary in his family and attaches to his lineal representative of to-day. One descendant of Scrimgeour was made constable of Dundee by William Wallace. Another bore the royal standard at Harlaw and fell in the battle. Sir Alan Durward (i.e. door-ward), whose castle was situated near the Loch of Lintrathen, held high office under Alexander II, married one of the king's daughters, was one of the regents during the minority of Alexander III, and in 1264 was one of the leading generals at Largs. In the thirteenth century the Montealts or Mowats, who had property near Fern, were an important Forfarshire family. One of them went to Norway as a witness of the marriage of Princess Margaret to King Eric.

Another family of historical interest is that of the Bethunes, Betons, or Beatons, whose descent has been traced to the time of William the Lion. James Beaton, Abbot of Arbroath and subsequently Archbishop of St Andrews, was succeeded in both offices by Cardinal David Beaton. The latter owned various lands in Forfarshire, and built Melgund Castle.

James Hallyburton of Pitcur Castle was Provost of Dundee in the time of Mary of Guise. He was a Protestant leader and protected Paul Methven (a noted reforming preacher of Dundee) against the royal mandates. When Mary was being besieged in Leith, Hallyburton led a contingent from his town, and was killed in a sortie from Leith. His son, provost of Dundee for thirty-three years, held many high offices during the reign of Mary, and was one of the committee who brought about her demission of the crown.

Sir Thomas Lyon, the Master of Glamis, was conspicuous in the Raid of Ruthven. It was he who exclaimed, "Better bairns greet than bearded men!" and at his death James is said to have remarked that the boldest and hardiest man in his kingdom was dead.

The "Great" Marquis of Montrose is one of the most noted of the sons of Forfarshire. At first a Covenanter,

Admiral Duncan

he went over to the king's side in 1642, and for some time was the terror alike of his native shire and of the whole country. His brilliant career and its sad termination have been the theme of many a writer.

Few men have been more eulogized by friends and defamed by foes than John Graham of Claverhouse, created Viscount Dundee by James VII. In Scott's *Old Mortality* he is depicted as a stern yet chivalrous soldier. Another native of Forfarshire shared with Claverhouse the epithet of "bloody"—Sir George Mackenzie. Born in Dundee (1636), he was educated at St Andrews and Bourges. Becoming King's Advocate, he earned the undying hatred of the Covenanters for his stern severity as prosecutor. But he was a lover of literature, and his best title to fame is the foundation of the Advocates' Library in Edinburgh.

Admiral Duncan was raised to the peerage as Viscount Duncan for his naval exploits, and especially for his victory at Camperdown. "The rapidity of his decision, the justice of his glance, were equal to those of Nelson himself."

Sir William Chalmers served with distinction under Wellington in the Peninsula and at Waterloo, and was knighted for his services.

From 1762 to 1790 the Forfar and Fife District of Burghs was represented in Parliament by George Dempster of Dunnichen. He was nicknamed "Honest George" because of the deep interest he took in the welfare of his native county. Equally distinguished in the beginning of the nineteenth century was George Kinloch. He presided over a monster meeting in Dundee to protest against the "Peterloo massacre" and so roused the ire

of the authorities that he was outlawed and was in exile
for three years. Pardoned by George IV, at the personal
request of his daughter, Kinloch returned to his native

George Wishart

town of Dundee, and sat as its first representative in the
Reformed Parliament. William, Lord Panmure, is known
as "the father of reform in Scotland," and still more widely
famous was his son, Fox Maule, Secretary of War. Of

another statesman, Joseph Hume, Montrose, his native town, is justly proud. In India he acquired so profound a knowledge of Indian languages that he was appointed Government Interpreter and Commissary-General. On his return home he became a great radical reformer in Parliament and won for himself the esteem and confidence of the whole nation.

One of the outstanding figures in the Reformation struggle was George Wishart, a native of Forfarshire. Teacher of Greek and preacher of the reformed doctrines, he became a marked man and was burned in St Andrews in 1546. Another Forfarshire martyr, Walter Mill, long parish minister of Lunan, was burned in St Andrews at the age of eighty-two. In later times the Rev. James Guthrie, son of the Laird of Guthrie, was executed for having promoted the Western Remonstrance, and for having denied the authority of the king in ecclesiastical matters. Dr Alexander Leighton, Usan, was one of Laud's victims, being mutilated in 1630 by order of the Star Chamber for a virulent libel against the bishops. His son, Robert Leighton, Bishop of Dunblane and afterwards Archbishop of Glasgow, proved himself a most amiable and broad-minded churchman.

The great names of Andrew Melville, James Melville his nephew, and Erskine of Dun, all closely connected with Montrose, take us back to the sixteenth century. With the exception of Knox, no preacher had more influence in ecclesiastical affairs than Andrew Melville. Melville was renowned as a scholar. He was Professor of Latin in Geneva, and, on returning to Scotland, became Principal of Glasgow University and then of St Mary's College, St Andrews, where at the same time his nephew was Professor of Oriental Languages. Andrew was for four years imprisoned in the Tower of London, and died in exile at Sedan in 1622. John Erskine of Dun, a member of a well-known family, returned from the Continent imbued with reformation principles, yet so highly respected by all that he frequently played the *rôle* of mediator between Catholics and Protestants. It was he who brought to Montrose Marsilliers, the first teacher of Greek in Scotland. After the Reformation in 1560, he was appointed Superintendent of Angus and Mearns. Without sacrifice of his integrity, he managed to retain the confidence of the Reformers and the favour of James VI.

Thomas Erskine of Linlathen, near Dundee, advocate and theologian, a friend of Carlyle, Dean Stanley, and F. D. Maurice, was an accomplished scholar, but was still more potent in stimulating the religious life of his times. Dr Thomas Guthrie, a native of Brechin, had his first ministerial charge in Arbirlot, though it was in Edinburgh that he won his fame. He was one of the Disruption leaders; and his deep interest in the poor manifested itself in the founding of the Edinburgh Original Ragged School.

Hector Boece may head the list of scholars and men of letters. Born in Dundee in 1465, he was appointed first Principal of the University of Aberdeen. His Latin *History of Scotland* is picturesque but largely legendary. James, John, and Robert Wedderburn, sons of a Dundee merchant, disciples of Wishart, and students of St Andrews, were contributors to the *Gude and Godly Ballates*, sometimes known as the Dundee Psalms. Sir Peter Young of Seaton, a contemporary of the Wedderburns, and like them the son of a Dundee burgess, was tutor, under Buchanan, to James VI, who knighted him for his services.

Amongst the many sons of Angus who have written of their native county are John Ochterlony or Auchterlony, who penned an interesting and trustworthy local history about 1682; James Thomson, "father of our local archaeological literature," the author of a number of valuable historical works on the county; and Alexander J. Warden, whose five monumental volumes are the result of years of patient research.

The Bards of Angus and Mearns proves this part of Scotland to be a nest of singing birds. One of these, Alexander Ross, schoolmaster of Lochlee, but a native of Aberdeenshire, honoured alike as a teacher and a poet, wrote *Helenore, or The Fortunate Shepherdess*, and other poems in the Scottish dialect. Amongst his popular songs are "The rock and the wee pickle tow," "To the beggin' we will go," and "Woo'd and married and a'." He died in 1784. Other two poets, natives of Dundee, are Robert Nicoll, "a second Burns," and William Thom, author of "The Mitherless Bairn."

Dr Thomas Dick, a native of Dundee, did much to popularise science, especially astronomy, by his books and his lectures. He died at Broughty Ferry in 1857.

James Tytler, son of a minister of Fern, wrote songs in his native Doric, and became first editor of the

Encyclopædia Britannica. One of his contributors, the son of a crofter at Logie-Pert, near Montrose, was James Mill, the historian of British India (1806–1818) and father of John Stuart Mill. Patrick Chalmers, M.P., of Aldbar, was noted alike for his philanthropy and his wide anti-quarian researches. His *Sculptured Stones of Angus* is a work of much value. Dr John Jamieson, author of *An Etymological Dictionary of the Scottish Language*, though not a native of the county, was long closely associated with it.

In science some natives of Forfarshire have gained not a little distinction. Dr Patrick Blair, born in Dundee, who was imprisoned after 1715 as a Jacobite, was a noted physician and botanist. Colonel William Patterson, son of a gardener in the parish of Kinnettles, made valuable contributions to the science of natural history. Towards the close of the eighteenth century he travelled in Africa, and from the Cape penetrated farther into the heart of the continent than any other European had then done. For many years he was Lieutenant-Governor of New South Wales. William Gardiner, also of humble birth, rose to fame as a botanist, and published a number of works, amongst which was *The Flora of Forfarshire.* Sir Charles Lyell, the eminent geologist, whose works revolutionised his science, was born in 1797 at Kinnordy, near Kirriemuir. In 1859 James Bowman Lindsay, a native of Carmyllie, held wireless communication across the Tay at Glencarse, where the river is half a mile wide. It was he also who originated the idea of a submarine cable to America. Lord Brougham said of another Dundonian, Sir James Ivory, that he was the greatest mathematician since Sir Isaac Newton.

Sir Charles Lyell

From the crowd of names connected with improvements in spinning and weaving appliances, it can scarcely be regarded as invidious to select that of the late James Carmichael of engineering fame.

26. THE CHIEF TOWNS AND VILLAGES OF FORFARSHIRE.

(The figures in brackets after each name give the population in 1911, and those at the end of each section are references to the pages in the text.)

Aberlemno, a parish (728) and village about six miles north-east of Forfar. Balgavies Loch was at one time dredged for marl. A greyish sandstone is plentiful in the district. Interesting ruins are Melgund and Flemington Castles, while Aldbar Castle, Balgavies, and Carsegowrie are ancient but still inhabited houses. Two sculptured stones, one in the churchyard, are objects of interest. (pp. 49, 83.)

Arbirlot, a coast-parish (840) and village 2¾ miles west by south of Arbroath, a picturesque and secluded spot. Kelly Castle, an ancient pile, stands in a wooded glen. (pp. 49, 99, 145.)

Arbroath (20,648). The royal burgh of Arbroath is the second town of Forfarshire in population and industrial importance. Its chief industries are the manufacture of jute and linen goods and fishing. It has also a shipbuilding yard, bleachfields, tanneries, engineering and chemical works. The chief attractions of the place are its high antiquity, its fine situation and bracing air, the grand scenery of its cliffs, and its noble abbey now in picturesque decay. There is a fine golf course at Elliot, the railway junction 1¾ miles to the south-west for the Carmyllie stone quarries, celebrated for "Arbroath paving-stones." (pp. 18, 34,

35, 36, 37, 40, 41, 42, 46, 49, 54, 61, 64, 66, 69, 71, 75, 80, 84, 85, 95, 100, 106, 116, 123, 125, 128, 132, 133, 135, 136, 138, 139.)

Auchmithie, a picturesque fishing village in the parish of St Vigeans, 3½ miles north-north-east of Arbroath. (pp. 38, 71.)

Auchterhouse, a parish (629) and village of south-west Forfarshire. The parish church is a good specimen of early church architecture in Scotland. The ground rises in the north and north-west to the Sidlaw Hills. High up in an excellent situation is the Sidlaw Sanatorium for the treatment of consumption. "Weems" have been discovered in the hills. (pp. 87, 91, 114.)

Baldovan, a village in the parish of Mains and Strathmartin, three miles north-west of Dundee, has an asylum for imbecile children, erected in 1854 by the late Sir John Ogilvy, the first and for long the only institution of its kind in Scotland.

Barnhill, a residential suburb of Dundee one mile north of Broughty Ferry.

Barry, a parish (4933) and village on the south-east coast. The south-eastern point of the parish is Buddon Ness. The links immediately to the north of this are utilised by the War Department for military camps and big gun practice. (pp. 13, 34, 40, 43, 50, 60.)

Brechin (8439). The city of Brechin is one of the most ancient towns of Forfarshire, having been "dedicated to the Lord" in the tenth century. Even earlier it was probably an ecclesiastical centre, and it certainly has been so since. Indeed it is entitled to the designation of "city," because of its cathedral. Its situation on the left bank of the South Esk, 8½ miles west-north-west of Montrose, is exceedingly picturesque, especially when viewed from the south. The modern town is an important manufacturing centre. Its chief products are osnaburgs, brown linen, and sailcloth; while bleaching, brewing, distilling and the making of machinery are extensively carried on. (pp. 16, 54, 61, 62, 81, 82, 83, 91, 102, 103, 115, 116, 122, 123, 127, 128, 131, 132, 133, 135, 136, 138, 139, 145.)

Broughty Ferry (11,059). In 1498 Andrew, third Lord Gray, built the castle of Burgh-Tay, now Broughty, on the small rocky peninsula 3½ miles east of Dundee. The village consisted

Carnoustie Sands

of little more than a few fishermen's huts for two or three centuries. Even in 1792 its population was only 230. To-day it is the third town of Forfarshire. This remarkable increase is due to its proximity to Dundee, of which it is the most important residential suburb. Its ferry, which connects Forfarshire and Fifeshire, was a vital link between north and south before the building of the Tay Bridge. (pp. 34, 40, 46, 53, 54, 71, 72, 85, 86, 108, 117, 118, 135, 146.)

Carmyllie, a parish (847) and village, with very fine quarries, about seven miles north-west of Arbroath. (pp. 70, 147.)

Carnoustie (5358), a coast town about 11 miles east-north-east of Dundee, of which it is largely a residential suburb, has important linen mills, vitriol works, and a brick and tile yard. Its sea-bathing facilities and its fine golf links have made it a popular summer resort. (pp. 13, 34, 53, 54, 83, 122.)

Clova, a highland hamlet in Glen Clova. The Kirkton is 15 miles north-west of Kirriemuir, near the entrances to passes leading through the Grampians to Aberdeenshire. (pp. 67, 87, 111.)

Craig, a parish (1883) and village directly south-west of Montrose, $1\frac{1}{2}$ mile distant. The parish contains Kirkton of Craig, and the fishing villages of Ferryden and Usan or Ulysses Haven. The most important mansions are Rossie Castle, Dunninald House, and Usan House. Rossie Reformatory was established in 1857. (p. 99.)

Dundee (165,006). The name of Dundee, the third town in Scotland, was in older times spelt in various ways—Donde, Dondie, and Dondei. The early history of the city is very obscure. Malcolm Canmore and some of his descendants appear to have done it honour. In the early thirteenth century it seems to have been the most wealthy and influential town in Scotland. Again and again, from Edward I to General Monk, it suffered siege and pillage. About 1650 it was second only to Edinburgh, but so disastrous was its treatment by Monk that it fell irretrievably behind. Prosperity returned in the eighteenth century with the linen industry, and during the succeeding century its growth was phenomenal. (pp. 2, 14, 20, 33, 34, 39, 41, 43, 44, 46, 47, 49, 53, 54, 61, 62, 63, 64, 66, 67, 69, 70, 71, 72, 73, 74, 75, 79, 83, 84, 85, 86, 87, 89, 95, 105, 108, 117, 129, 132, 133, 135, 136, 138, 139, 140, 142, 145, 146, 147.)

Dunnichen. This Forfarshire parish (1098) contains the villages of Dunnichen and Letham. The Mire of Dunnichen,

50 acres, has been drained and cultivated. Vestiges of a Pictish fort have been all but obliterated, and another ancient fortification is still pointed out on Dunbarrow Hill. Numerous stone-covered graves and a round sepulchral knoll have disclosed urns and human bones, believed to be memorials of those who fell in the great battle of Nechtan's Mere. (pp. 49, 50, 60, 82, 98, 142.)

Edzell, a pretty village about six miles north by west of Brechin, is a favourite summer resort. The parish (878) stretches north and south for nearly 12 miles, and east and west for nearly six. (pp. 17, 67, 68, 113, 115, 135, 138.)

Forfar (10,849), the county town, is a royal and parliamentary burgh, 14 miles north-north-east of Dundee, and a place of great antiquity. It was a favourite residence of Malcolm Canmore and Queen Margaret; and later monarchs William the Lion, Alexander II, and Robert II are said to have held parliaments within its walls. Its modern importance is a mere shadow of its ancient standing. It manufactures jute and linen goods. The "sutors" of Forfar were in times now long gone by famous for the making of wooden brogues or shoes. (pp. 54, 61, 62, 81, 90, 95, 97, 103, 106, 115, 129, 132, 133, 135, 136, 138, 139.)

Friockheim, a village in Kirkden parish (1337), began to be important in 1839 as a station on the main line between Forfar and Arbroath.

Glamis, a village and parish about six miles south-west of Forfar, contains the historic Glamis Castle (pp. 114, 115, 121, 122).

Inverarity, a village and parish (861) in the Eastern Sidlaws, has interesting antiquarian remains. (p. 49.)

Inverkeillor is a village and parish (1376) on the coast six miles north by east of Arbroath. (pp. 49, 83.)

Kettins. The village stands in a parish (689) of the same name about two miles east-south-east of Coupar Angus. Picts' houses, i.e. weems, have been discovered in the neighbourhood. Hallyburton, Lintrose, Baldowrie, and Bandirran are the chief mansion houses. (pp. 43, 99.)

Glamis Castle

Kirriemuir (3776) is a police burgh situated on rising ground five miles west-north-west of Forfar. It commands a magnificent view of Strathmore, and backed by the long line of the Forfarshire Grampians is itself a notable object in the landscape. Its handloom weavers gave the town an early reputation in the manufacture of brown linen, which it has by no means lost since the introduction of the power-loom. Hence the name of "Thrums" by which one of its most gifted sons, Mr J. M. Barrie, has designated his native-place in his novels. Until 1875 Kirriemuir was a burgh of barony under the Earl of Home. (pp. 54, 61, 62, 67, 91, 99, 106, 120, 131, 133, 135, 136, 138.)

Lochee (14,845) forms part of the parliamentary burgh of Dundee. It is situated to the north-west of the ridge connecting the Law Hill and Balgay Park, a finely wooded and picturesque eminence. Hand-loom weaving of coarse linen fabrics was its initial industry, to which was soon added bleaching, and then spinning, dyeing, printing, and calendering. The chief factory of the town is that of Messrs Cox Brothers, one of the largest in the world. Its conspicuous chimney-stalk is 282 feet high. As many as 5000 hands have been employed in this gigantic work; while 24,000,000 yards of sacking, and 14,000,000 yards of other fabrics have been turned out in one year. (pp. 135, 146.)

Logie-Pert, a parish (1002) and village about $4\frac{3}{4}$ miles north-west of Montrose. The parish contains the village of Craigo. (pp. 60, 147.)

Monifieth (3098), a village with fine golf-links, $5\frac{3}{4}$ miles east-north-east of Dundee, has many villas but is not entirely residential, there being a jute mill and a foundry and machine works. (pp. 13, 40, 53, 60, 95, 99, 135.)

Monikie. The parish (1184) contains the villages of Monikie, Craigton, Guildy, and Newbigging. There are excellent building stone and pavement quarries at Pitairlie. Affleck

Castle is one of the finest buildings of its kind in the county; and the Panmure Monument is a conspicuous object. (pp. 49, 118.)

Montrose (10,973). A fine beach and extensive links border the ocean between the two rivers and render Montrose an attractive seaside resort. Bricks and tiles are manufactured at Dryleys and Puggieston. Two miles north-west of Montrose is Sunnyside Asylum, one of the best establishments of its kind in the country. Hillside is a residential suburb with fine villas. The town of Montrose is a royal burgh, a seaport, and a manufacturing centre, and seems to have been in existence as early as the tenth century. A suspension bridge and a railway viaduct span the South Esk. Flax-spinning, rope-works, tanneries, machine-making establishments, breweries, starchworks, soap-works, an artificial manure and chemical work, and boat-building are the chief industries of the town and district. The harbour is important, and Montrose is the headquarters of the local fishery district. (pp. 13, 34, 40, 43, 47, 54, 61, 62, 64, 69, 70, 71, 72, 73, 75, 79, 83, 87, 89, 106, 116, 127, 133, 135, 136, 138, 139, 144, 145.)

Newtyle is a village that owed its origin in 1831 to the Dundee and Newtyle railway. Kinpurnie Hill (1134 feet) with its disused observatory is a far-seen object and commands an excellent view. The Glack of Newtyle is a pass behind the village, through which run road and railway. (pp. 114, 133, 135.)

St Vigeans is a small village a little over a mile from Arbroath, interesting for its great antiquity. (p. 99.)

Tannadice is a village on the South Esk, seven miles north-east of Forfar.

Tarfside is the chief village in Glen Esk, near the junction of the Tarf and the North Esk. In the vicinity are Migvie or Rowan Hill with a monumental cairn, the ruins of Invermark Castle, and the shooting lodge of Invermark.

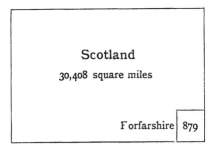

Fig. 1. Area of Forfarshire (879 sq. miles) compared
with that of Scotland

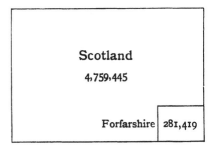

Fig. 2. The Population of Forfarshire compared with
that of Scotland (1911)

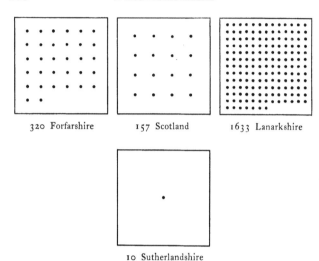

320 Forfarshire   157 Scotland   1633 Lanarkshire

10 Sutherlandshire

Fig. 3. Comparative density of Population to the square mile in certain Scottish counties (1911)

(Each dot represents 10 persons)

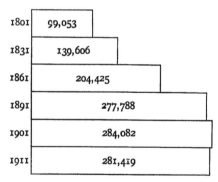

1801	99,053
1831	139,606
1861	204,425
1891	277,788
1901	284,082
1911	281,419

Fig. 4. Diagram showing increase and decrease of Population in Forfarshire since 1801

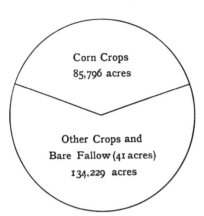

Fig. 5. Proportionate area under Corn Crops compared
with that of other cultivated land in Forfarshire in 1911

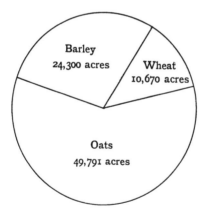

Fig. 6. Proportionate area of chief Cereals in
Forfarshire in 1911

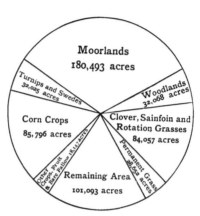

Fig. 7. Proportionate areas of land in Forfarshire in 1911

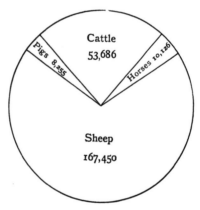

Fig. 8. Proportionate numbers of chief Live Stock in
Forfarshire in 1911